walkermaths 1.1
Exploring Data

Charlotte Walker and Victoria Walker

Walker Maths: 1.1 Exploring Data
1st Edition
Charlotte Walker
Victoria Walker

Cover design: Cheryl Smith, Macarn Design
Text design: Cheryl Smith, Macarn Design
Production controller: Siew Han Ong

Any URLs contained in this publication were checked for currency during the production process. Note, however, that the publisher cannot vouch for the ongoing currency of URLs.

Acknowledgements
The authors and publisher wish to thank the following people and organisations for permission to use the following resources in this workbook.
- Carey Knox for the Gecko data and cover picture.
- Murray Williams for the Whio data.
- Auckland Zoo for the visitor data.
- All other images are courtesy of Shutterstock and iStock.

We also wish to acknowledge the trusted kindred spirits throughout the country for your help in preparing this title.

For product information and technology assistance,
in Australia call **1300 790 853**;
in New Zealand call **0800 449 725**

For permission to use material from this text or product, please email **aust.permissions@cengage.com**

National Library of New Zealand Cataloguing-in-Publication Data
A catalogue record for this book is available from the National Library of New Zealand.

978 01 7047743 7

Cengage Learning Australia
Level 5, 80 Dorcas Street
Southbank VIC 3006 Australia

Printed in China by 1010 Printing International Limited.
3 4 5 6 7 27 26 25 24

This icon appears throughout the workbook. It indicates that there is a worksheet available which has been collated from the original Level 1 WalkerMaths series. The worksheets can be accessed via the Teacher Resource for this workbook (available for purchase from nz.sales@cengage.com). These worksheets are an additional resource which can be used to support your students throughout the teaching of this standard.

CONTENTS

Statistical enquiry cycle

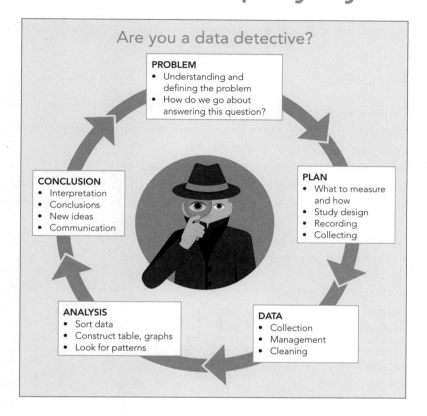

This book explains how to carry out four types of statistical investigations:

Relationships

You will investigate whether for **one group** there is a relationship between **two numeric variables**.

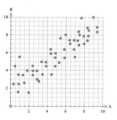

Comparisons

You will investigate whether there is a difference between **two groups** for **one numeric variable**.

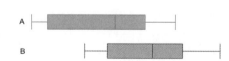

Time series

You will investigate how **one variable** changes over **time** and make a prediction.

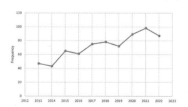

Probabilities

You will carry out a **probability investigation** and describe your observed probabilities in context.

Statistical concepts

- It's important to understand some terms and concepts that are important in Statistics.

Census vs sample

A census compared with a sample

Census
- You collect data from **every member of the population**.
- You get very accurate information.
- However, it is often impossible and usually very expensive to do.
- It is best to do a census if the answer to the question **really matters**.
- In New Zealand, a census of the population takes place every five years.

Sample
- You collect data from **just some of the population**.
- The information you get will not be as accurate.
- However, it is much easier and cheaper to take a sample.
- It's important that the sample is selected *fairly*.

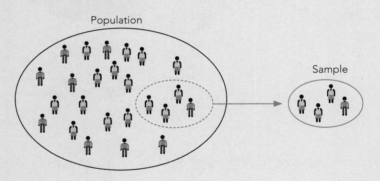

Which would be more appropriate to answer these questions and why?

	Question	Census or sample?	Why?
1	What is the most popular cellphone brand?		
2	Who will be the next New Zealand prime minister?		
3	What do New Zealanders think about a new crisp flavour?		
4	Do New Zealanders want the flag changed?		
5	Do New Zealanders want to become part of Australia?		

 ISBN: 9780170477437

Types of variables

There are three type of variables:

Categorical variables
These are descriptions or names.
Data recorded about each person/thing: **words**.
Typical question starts with 'What …'
Examples: colour, type of pet, favourite something.

Discrete variables
These are **numbers** which are the result of **counting**.
Data recorded about each person/thing: **whole numbers**.
Typical question starts with 'How many …'
Examples: number of pets, number of T-shirts.

Continuous variables
These **numbers** are the result of **measuring**.
They can be **fractions** or **decimals**.
Typical question starts with 'How long, heavy, etc. …'
Examples: height, mass, distance, time.

Highlight the correct variable type for each of these.

1	Eye colour	Categorical	Discrete	Continuous
2	Time it takes you to get to your next class	Categorical	Discrete	Continuous
3	Favourite chocolate bar	Categorical	Discrete	Continuous
4	Number of cousins you have	Categorical	Discrete	Continuous
5	Distance to the nearest supermarket	Categorical	Discrete	Continuous
6	Number of pizza shops in your neighbourhood	Categorical	Discrete	Continuous

Here is some data that has been collected.

	Question	Remi	Lucy	Gia
A	How many pets do you have?	3	1	0
B	What is your favourite chocolate?	White	Mint	Milk
C	Will you buy your lunch today?	No	No	Yes
D	How long is your driveway?	13 m	101 m	4.5 m
E	How many bedrooms are in your house?	3	6	4
F	What is the mass of your pencil case?	0 g	236 g	109 g

7   Which questions have answers that are categorical?   _____

8   Which questions have answers that are discrete?   _____

9   Which questions have answers that are continuous?   _____

Data display

- There are different ways of displaying data. Which is most appropriate depends on the **type** of variable:

Categorical variable	Discrete variable	Continuous variable
Tally chart Pictograph Dot plot Bar graph Strip graph Pie graph	Tally chart Pictograph Dot plot Bar graph Strip graph Line graph/time series Scatter plot Box plot	Histogram Line graph/time series Scatter plot Box plot **If rounded** Dot plot

Types of variables required for each graph

Graph	Picture	Variables required
Scatter plot		Two numerical variables (continuous or discrete).
Comparative dot plots		One numeric variable (continuous or discrete) for two groups.
Comparative box plots		
Time series		One **time** variable and one numeric variable (continuous or discrete).

     ISBN: 9780170477437

Investigative questions

- There are **three** types of investigative questions.

Summary questions: investigate **one variable** at a time. This could be categorical or numeric.

    Examples: What is your favourite fruit?   **One** variable — **fruit**.

                     How many pairs of socks do you have?   **One** variable — **pairs of socks**.

Comparative questions: investigate **one numeric variable** for **two groups**.

    Example: Did the boys tend to get higher scores than girls in the driving test?

        One numeric variable — **driving score**. **Two** groups — **girls** and **boys**.

Relationship questions: compare **two numeric variables** for **one group**.

    Example: Is there a relationship between height and arm span in Year 11 students?

        Two numeric variables — **height** and **arm span**. **One** group — Year 11 students.

For each of these, highlight the type of question asked and write down an appropriate graph.

	Question	Type of question	Type of graph
1	What is the most common colour of schoolbag?	Summary Comparative Relationship	
2	Does the length of time spent gaming each day have an effect on the number of NCEA credits earned?	Summary Comparative Relationship	
3	How much do students spend on stationery at the start of the year?	Summary Comparative Relationship	
4	Do blond people tend to have a faster reaction time than brunettes?	Summary Comparative Relationship	
5	How many movies did students see in the last year?	Summary Comparative Relationship	
6	Do electric cars tend to be more expensive than cars that run on petrol?	Summary Comparative Relationship	
7	How far do students live from the nearest supermarket?	Summary Comparative Relationship	
8	Does height have an influence on the time taken for students to run 100 m?	Summary Comparative Relationship	

Sampling and bias

It's important to consider whether a sample is likely to be representative of the population.

Sample size
- Each sample needs to be large enough to ensure that it is representative of the population.
- The larger the sample size, the more closely it is likely to reflect the population.

Sampling method
- A random sample means that the sample is much more likely to be representative of the population.
- Bias occurs when some members of the population are more likely than others to be selected for the sample, so the sample does not truly represent the population.
- Sampling randomly is more likely to result in data that is representative of the population.

1 a Count the number of geckos and birds in each sample.

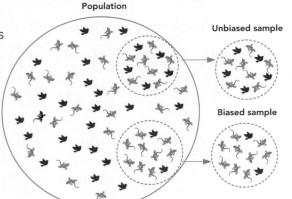

Geckos _____

Birds _____

Geckos _____

Birds _____

b The second sample is biased because the numbers in the sample are/are not proportional to the population.

2 Let the population be all students attending a school. State whether the following samples are likely to be biased or not, and explain why. Some have been done for you.

	Question	Biased/Unbiased	Reason
a	Every tenth student on the school roll.	~~Biased~~/Unbiased	Every student in the school has an equal chance of being selected.
b	Those who put suggestions into a box for ideas for the school ball.	Biased/~~Unbiased~~	This is a 'self-selected' sample because each student decides whether or not they will be sampled.
c	Students seated near the tuckshop at lunchtime.	Biased/Unbiased	
d	Class 10Wl.	Biased/Unbiased	
e	Randomly selected students from the school roll.	Biased/Unbiased	

ISBN: 9780170477437

 # Variation

- The **values** of variables that we record can **change**, depending on how, when or where the data was collected. This is known as **variation**.
- It is important to consider the variation in the data for your investigation, and try to minimise it.
- Some sources of variation will be able to be managed. Some won't be.

Types of variation

- As part of your investigation, you need to identify variation, its type and how you plan to manage it.
- This should be documented in your plan and should also be reflected on after gathering your data.

Occasion-to-occasion variation

This is when the same variable for an individual is measured at different times.
E.g. your heart rate is likely to be different at different times of the day.

Measurement variation

This occurs when there is a difference between the actual value and the recorded value.
This could be due to any of the following:
1 An inaccurate measuring device.
2 Inaccurate measurement caused by human error, which could be:
          transcriptional error, e.g. writing 89 instead of 98
   or   estimation error, e.g. reading 98 mm on a ruler instead of 99 mm.
3 Inaccurate measurement caused by the subject.
   E.g. an animal might be wriggling while it is being measured.

Induced variation

This occurs when there is variation in the values of the variable caused by:
1 The experimental process.
   E.g. the conditions for each individual in an experiment were not identical. One plant might grow taller than another because it was watered more regularly, had different soil, or its position meant it received more sunlight than another.
2 The measuring conditions or process.
   E.g. measuring of animals may cause damage or trauma, which affects their growth.

Situational variation

This occurs when there is variation in values of the variable because of previous occurrences or conditions.
E.g. when an animal is being weighed, its mass will be affected by how recently it has eaten; the time a person takes to run 100 m is likely to be affected by how tired they are.
Sometimes this variation can be managed by good planning.

Sampling variation

- Every sample from a population will be different.
- The object is to take a sample that is representative of the population.
  - ∴ 1 The sample needs to be large enough.
  -    2 The sample must not be biased.

If samples are representative of the population, then while sample statistics are likely to be different for each one, they should still lead to the same conclusion.

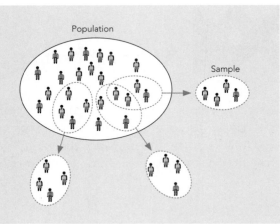

Variation for different types of data

Primary data — data you have collected yourself

- Describe in detail what variation you anticipated or encountered.
- Identify what type of variation each is.
- Explain how you managed the variation.

Example: Puppies are being weighed to track their growth.

What could cause variation	Type of variation	How could it be managed?
The puppy might not stay still while being weighed.	Measurement variation	Scales must be set to zero and the puppy should be placed in a box to limit movement.
The puppy might be wet.	Situational variation	The puppy should be dry.
The time of day.	Occasion-to-occasion variation	Weigh the puppy at the same time each day.

Secondary data — data you have been given

- Describe at least two assumptions regarding variation that may have occured during collection.
- Describe at least two assumptions you have made about its collection.
- Consider how you would have managed the variation if you collected it.
- Identify the types of variation.

Example: The time taken for a child to complete a jigsaw puzzle.

What could cause variation	Type of variation	How it could have been managed
The child might have seen the puzzle before.	Situational variation	Making sure it's the first time the child has seen this puzzle.
The pieces might not have been mixed up properly.	Induced variation	Ensuring that the pieces are thoroughly mixed before the child sees them.
Mistakes in recording the data. E.g. 1 min and 10 seconds recorded as 1.1 min.	Measurement variation	Have somebody checking the stopwatch and the recorded time.

ISBN: 9780170477437

Answer the following questions.

1 When measuring students' arm spans, highlight the things that need to be considered in order to minimise variation.

Positions of fingers	What they have for breakfast	Age	Device used for measurement
Height	Hairstyle	Angle between arms and torso	Bicep size

2 The lengths of time some children took to tie their shoelaces have been measured. Name the types of variation listed below, and suggest a way of managing each.

	Possible causes of variation	Type of variation	How could it be managed?
a	The temperature at the time the measurement is being done. This might affect the dexterity of the children's fingers.		
b	Stopping a stopwatch at the exact time of finishing can be challenging and will depend on the timer's reaction times.		
c	The time of day at which measurements are made. Tiredness and concentration are unlikely to be constant throughout the day.		

3 The tail lengths for a colony of geckos are recorded each year. Name the types of variation listed below, and suggest a way of managing each.

	Assumptions about variation	Type of variation	How it could have been managed
a	The geckos were kept still while they were being measured.		
b	The samples were representative of the population.		
c	That the process of measuring the tails of geckos does not affect their survival from one year to the next.		

Data check

- Whether you have collected or been given the data, it should be checked to ensure that every piece appears to be valid.
- This is known as **cleaning** the data.

Once an unusual piece of data is identified:
1 If possible, re-measure or re-count and check that the units are correct.
2 If it's an **obvious** mistake, then remove the piece of data and explain why it has been removed.

Examples:

1 Distance students travel each morning to get to school (to the nearest km):

2, 3, 1, **153**, 7, 4, 18, 12, 1, 21, 2, 6

> If possible, check with this student to determine just how far they travel.
> If not possible, then remove this piece. It is highly unlikely anyone travels 153 km to get to school.

2 Marks out of 10 for a test:

3, 10, 8, 3, **11**, 2, 7, 5, 0, 1

> If possible, check the test paper.
> If not possible, remove this piece. A mark greater than 10 is not possible.

Do not remove unusual points unless you have a good reason for doing so.

Highlight which points you would investigate, explain why and describe what you would do.

1 Heights of students (cm) in a Year 12 class:
153, 5′ 4″, 165, 179, 171, 1082, 159, 158, 149, 168

_____

_____

2 Number of hours slept last night:
7, 10, 7, 0, 3, 6, 29, 17, 4, 9, 10, 9, 6, 8, 21

_____

_____

3 The price of a motel room for one night:
$170, $110, $310, $200, $139, $165, $140, $17, $165, $160, $120

_____

_____

 ISBN: 9780170477437

 # Data analysis

There are two measures that we need to know in order to be able to discuss and compare distributions:

    1 Where is the **centre** of the data?
    2 How widely is the data **spread**?

Measures of centre

- There are **three** measures for the centre of the data.

Name	Calculation	Advantages	Disadvantages
Mean	$\dfrac{\text{the sum of all the data values}}{\text{the number of data values}}$	• Easy to calculate	• Distorted by values that are vastly different from the others.
Median	middle value	• Usually a good measure of centre	• Data must be put in order
Mode	value that occurs most frequently	• Very easy to find	• Very unreliable as measure of centre • Often there are two or none

- Avoid using the word 'average' because it can have different meanings.

Mean

- The numbers **do not need to be in order** for this calculation.
- Means are not always whole numbers, so **sensible rounding** may be needed.
- The mean is influenced by unusually large or small values.

$$\text{mean} = \frac{\textbf{sum of all the data values}}{\textbf{number of data values}}$$

Example:

0 3 8 2 1 8 5 7 64

Notice that 0 must be included in the calculation.

$$\text{Mean} = \frac{0 + 3 + 8 + 2 + 1 + 8 + 5 + 7 + 64}{9}$$

$$= 10.\dot{8}$$

There are 9 numbers in the data set.

The mean of this data set is 10.$\dot{8}$ or 10.9 (1 dp).

Notice what happens to the mean if the 64 is removed.

0 3 8 2 1 8 5 7

$$\text{Mean} = \frac{0 + 3 + 8 + 2 + 1 + 8 + 5 + 7}{8}$$

$$= 4.25$$

There are now 8 numbers in the data set.

The mean of this data set is 4.25.

The 64 was so much larger than the other values that it influenced the mean.

Median

- If there is an **odd number of values** in a data, the median is the **middle number**.
- If there is an **even number of values**, the median is **halfway between the two middle numbers** in the data set.
- Before you can calculate the median, you must **put the data in order**.

Examples:

1 A data set with an odd number of values

$$9 \quad 12 \quad 8 \quad 15 \quad 20 \quad 17 \quad 23 \quad 11 \quad 19$$

Put them **in order** before finding the median: **8 9 11 12 15 17 19 20 23**

The median of this data set = 15

This is the middle number.

2 A data set with an even number of values

$$0 \quad 1 \quad 2 \quad 4 \quad 8 \quad 9 \quad 14 \quad 15 \quad 17 \quad 18 \quad 18 \quad 21$$

The median for this data set = $\dfrac{9 + 14}{2}$ = 11.5

Mode

- The mode is the **most common value**.
- Sometimes there are **several modes**.
- If there are **three or more** numbers that occur equally often, we say there is **no mode**.

Examples:

1   ⑨ 16 21 8 5 ⑨ 12 19 4 12 ⑨ 15 3

The most common number is **9**: there are three of them.

The mode of this data set is 9

2   14 8 ⑬ 11 6 ⑩ 8 12 ⑬⑩ 9 15 4

Both **10** and **13** occur twice.

The modes are 10 and 13

3   ㉓ 26 ㉑ ㉚ 25 ㉑ 32 27 ㉚ ㉓ 29

There are three numbers that occur equally often: **21**, **23** and **30**.

If there are three or more **modes, we say the data is** polymodal.

     ISBN: 9780170477437

Calculate the mean, median and mode for these data sets. Round any calculations to 3 sf.

1   **0   1   1   4   6   7   12   15   18   21   24   25   29**

Mean = _____   Median = _____   Mode = _____

2   **9   12   12   15   18   19   22   25   27   27   30   39**

Mean = _____   Median = _____   Mode = _____

3   **0.3   0.8   0.9   1.7   1.9   2.0   2.6   2.6   2.8   2.9   2.9   3.1   3.5   3.5**

Mean = _____   Median = _____   Mode = _____

4   **21   23   23   25   25   27   28   28   29   30   32   93**

Mean = 32   Median = 27.5   Mode = 28

The mean and median are similar/different. Explain why.

_____

5   **51   53   53   56   58   59   59   60   61   65   66   68**

Mean = 59.1   Median = 59   Mode = 59

The mean and median are similar/different. Explain why.

_____

6   The following sets of numbers are in ascending order. Find the missing number.

a   5 5 6 [ ] 9 9 9 11   — the median is 7.5.

b   2 3 4 6 [ ] 9 9 9 10 12   — the mean is 7.

7   Here are three data sets. Find the missing numbers.

Clues:
1 They all have the same mean.
2 Set A has the same median as set B.
3 The median of set C is double the median of set B.

Set A	3	20	7
Set B	17		
Set C	15		

Measures of spread

Range
- The range is the **maximum** value **minus** the **minimum** value in the data set.
- Note: the range is a **single number**.
- Like the mean, the range is affected by unusually large or small values.
- The data does not need to be in order to calculate the range.
- The range is a measure of the **variability** of the data.

$$\text{range} = \text{maximum} - \text{minimum}$$

Quartiles
- The upper quartile (UQ) is the **median** of the **top half** of the data.
- The lower quartile (LQ) is the **median** of the **bottom half** of the data.
- The interquartile range (IQR) is also a measure of the **variability** of the data. It is usually considered to be a better measure of spread than range because it isn't affected by extreme values.

$$\text{interquartile range (IQR)} = \text{upper quartile (UQ)} - \text{lower quartile (LQ)}$$

Examples:

1 If the number of pieces of data is **odd**, **exclude** the median from the top and bottom halves of the data.

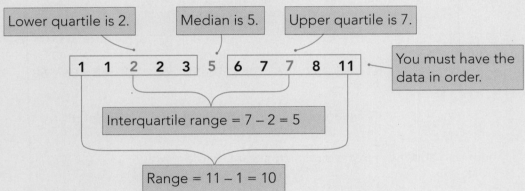

2 If the number of pieces of data is **even**, find the median values of the top and bottom halves of the data.

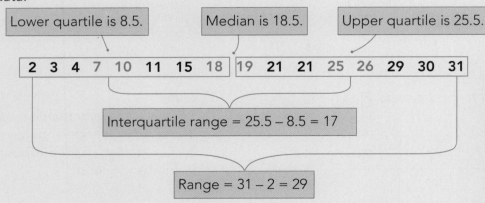

Ranges and interquartile ranges are used to measure the variability of groups of data.

ISBN: 9780170477437

Calculate the statistics for the following data sets.

1 7 9 14 18 19 25 27 29 33 37 40 47 49 49

Minimum = _____ LQ = _____ Median = _____ UQ = _____ Maximum = _____

Range = _____ Interquartile range = _____

2 15 19 23 28 31 35 38 41 44 47 51

Minimum = _____ LQ = _____ Median = _____ UQ = _____ Maximum = _____

Range = _____ Interquartile range = _____

3 2 3 3 4 5 6 7 9 9 10 12 12 15 16 18 20 20

Minimum = _____ LQ = _____ Median = _____ UQ = _____ Maximum = _____

Range = _____ Interquartile range = _____

4 9A and 9B took the same test. There were 10 questions. The table shows the results.

9A	6	7	6	6	5	7	4	5	6	5	5	6	5	7	5	4	8	6	4	5
9B	2	8	4	10	9	8	6	7	1	6	7	9	10	8	10	5	10	7	8	6

Calculate the mean, median, mode and range of these data sets.

9A																				
9B																				

9A Mean = _____ Median = _____ Mode = _____ Range = _____

9B Mean = _____ Median = _____ Mode = _____ Range = _____

Which class did better? _____

Explain your answer.

_____

_____

Relationships

- You need data for **two numeric variables** for **each** member of the population (or sample) and you analyse both at once to see if there is a relationship between them.
- A **scatter graph** has one variable on each axis and **each member** of the population or sample is represented by **one point**.
- The distribution of the points on the scatter graph is used to determine if there is a relationship (correlation) between the two variables.
- If a relationship exists, it can be used to make **predictions** or **estimates**.

For **relationship** investigations, we use samples to make an **inference** (suggestion) about the **correlation** (or otherwise) between **the two variables** in the population.

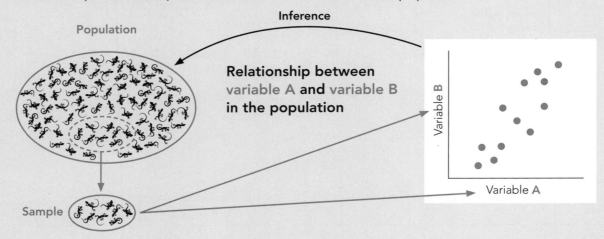

What's the point?

- The point is to compare **two numeric variables** for **one group**.
- The purpose of your statistical investigation should be clear and defined.
- Posing a question can help to clarify this.

Example: I wonder if there is a relationship between the time spent each day on devices (minutes) and the number of NCEA credits earned by Year 11 students at Paradise High School.

What do we need to think about?

What is the population? What population is your inference about?
Type of data — you must have **numerical** data for **two variables** for **one group**.

This can be discrete or rounded continuous data.

Sample size — you should have two pieces of data for **at least 30** individuals or items.

The larger the sample, the more likely it is to be representative of the population.

Where am I getting my data?

- If you are given it, where did it come from?
- If you are collecting it, you must describe how you did this.

Variation

Explain at least two different types of variation and how these could be managed.

Variable selection

- When undertaking a relationship investigation, **two numeric variables** should be selected.
- It's sensible to think about how or why one variable might affect another. This can help lead to interesting discussion.

There are several species of gecko in New Zealand. The table below shows data for two species. Elevation is the height above sea level that the gecko was found. SVL is the measurement of a reptile from snout to vent.

Species	Sex	Pattern	SVL (mm)	Mass (g)	Age	Elevation (m)
Orange spotted	Male	no orange	80	14	Adult	1510
Orange spotted	Female	orange spots	69	14.5	Juvenile	1519
Orange spotted	Male	spots	80	12.4	Adult	1500
Jewelled	Male	diamonds	74	11.2	Juvenile	1508
Jewelled	Male	barbed wire	75	10.6	Adult	1506
Jewelled	Female	diamonds	78	13.8	Adult	1488
Jewelled	Female	diamonds	71	10.1	Juvenile	1423
Orange spotted	Male	no orange	81	13.2	Adult	1404
Orange spotted	Male	orange spots	80	11.5	Adult	1377
Jewelled	Female	diamonds	67	10.9	Juvenile	1579
Orange spotted	Female	no orange	61	12.3	Juvenile	1581
Jewelled	Male	barbed wire	78	12.1	Adult	1583
Orange spotted	Male	orange spots	77	10.9	Adult	1583
Orange spotted	Female	orange spots	70	11.8	Adult	1582
Orange spotted	Female	orange spots	73	10.5	Adult	1512
Jewelled	Female	diamonds	76	11.9	Adult	1500

What variables from this data set would be appropriate to be used in a relationship investigation?

Variables	Reason for investigating these variables
Elevation and mass	Elevation could be related to mass, as food sources at different elevations could vary. This could affect how heavy the gecko is.

Scatter plots

- A scatter plot provides a visual representation of the **relationship** between **two numeric variables**.
- These variables can be for **discrete** or **rounded continuous** data.
- Each dot represents **two** pieces of data about **one** object, e.g. the height and mass of a person.

Understanding scatter plots

This graph shows the time spent studying and grades (%) of some New Zealand secondary school students.

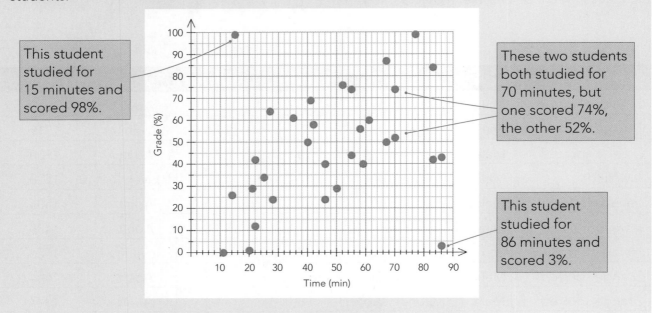

This student studied for 15 minutes and scored 98%.

These two students both studied for 70 minutes, but one scored 74%, the other 52%.

This student studied for 86 minutes and scored 3%.

1 The graph shows the bag masses and heights of some New Zealand secondary school students. Write down the coordinates of the **black** points and answer the questions.

Student	Coordinates	Description
a	(_____, _____)	This student is _____ cm tall and their bag mass is _____.
b	(_____, _____)	This student is _____ cm tall and their bag mass is _____.

c What mass was the bag of the tallest student?

_____

d How tall was the student with the bag of biggest mass?

_____

e One student didn't have a bag. How tall was this student?

_____

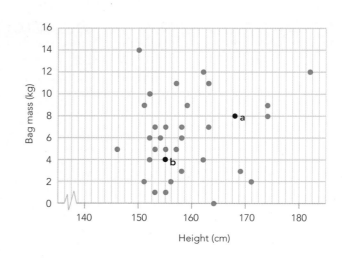

Describing features of scatter plots

Direction

Positive:

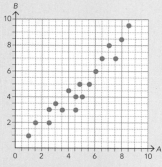

As A gets bigger, B gets **bigger**.

Negative:

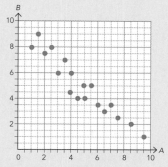

As A gets bigger, B gets **smaller**.

Strength

Strong:

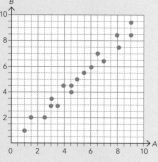

This means:
- The relationship is **strong**.

Moderate:

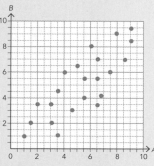

This means:
- The relationship is **moderate**.

Weak:

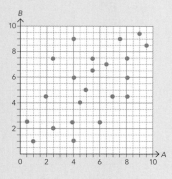

This means:
- The relationship is **weak**.

Trend

Linear trend:
Straight line

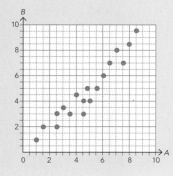

Non-linear trend:
Curved

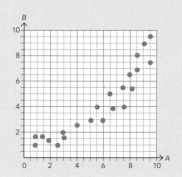

Identify the appropriate descriptions of these features.

1

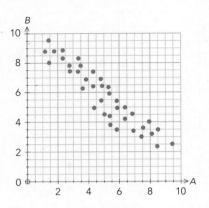

Direction   positive/negative

Strength   strong/moderate/weak

Trend   linear/non-linear

2

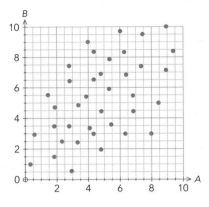

Direction   positive/negative

Strength   strong/moderate/weak

Trend   linear/non-linear

3

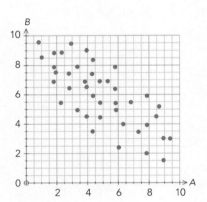

Direction   _____

Strength   _____

Trend   _____

4

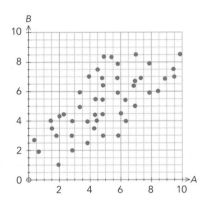

Direction   _____

Strength   _____

Trend   _____

5

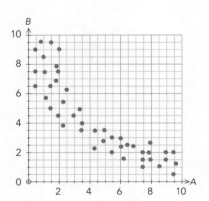

Direction   _____

Strength   _____

Trend   _____

6

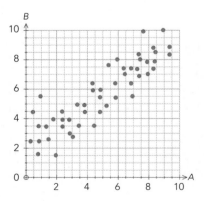

Direction   _____

Strength   _____

Trend   _____

Describing the meanings of features

- Just identifying the features isn't sufficient. You also need to **justify** your identification.
- In addition, for some features you need to **explain** what they mean in context.

Direction

Positive relationship:

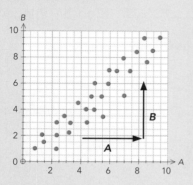

As A increases, B **increases**.

Negative relationship:

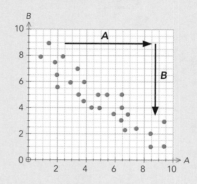

As A increases, B **decreases**.

Examples:

1 The relationship between students' left foot length and height.

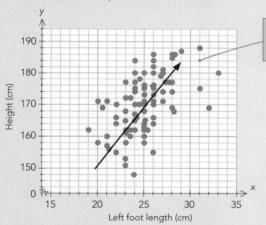

Data goes from the bottom left to the top right.

I notice that the relationship between the length of students' left feet and their heights is **positive**.

This means as the length of their left foot **increases**, their height also tends to **increase**.

2 The relationship between students' grades and the amount of time spent during the evening on social media.

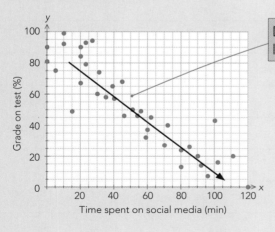

Data goes from the top left to bottom right.

I notice that the relationship between the time spent on social media and the grade in a test is **negative**.

This means as the time spent on devices **increases**, the grade tends to **decrease**.

Complete/write the statements about the direction of the relationships below.

1 This graph shows the relationship between the maximum daily temperature and the number of ice creams sold from a kiosk.

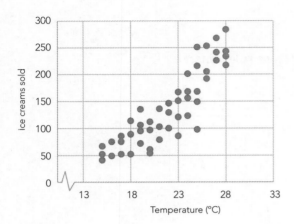

I notice that the relationship between

_____ and

_____ is positive/negative.

This means as the _____

increases, the _____

_____ tends to increase/decrease.

2 This graph shows the relationship between the ages of contestants and the time they took to complete a 5 km running race.

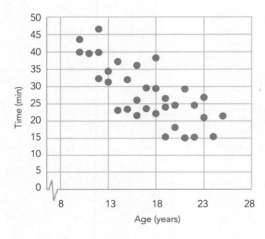

I notice that the relationship between

_____ and

_____ is

_____.

This means as the _____

increases, the _____

tends to _____.

3 This graph shows the relationship between the player's annual salary and number of NRL (rugby league) games played.

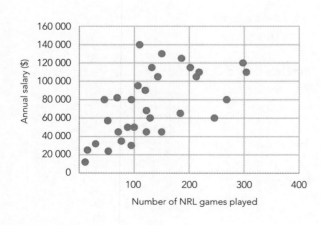

_____

_____

_____

_____

_____

_____

_____

_____

_____

ISBN: 9780170477437

Strength

- The amount of scatter tells you the strength of the relationship.

Strong:

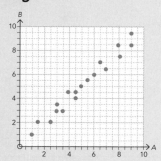

Moderate:

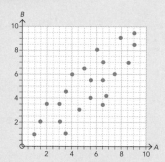

Weak:

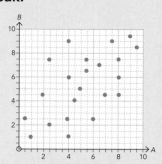

There is not a lot of scatter. This means the relationship is strong.

There is a moderate amount of scatter. This means there is a moderate relationship but it isn't strong.

There a lot of scatter. This means the relationship is weak.

Examples:

1 The relationship between the lengths of students' left feet and their heights.

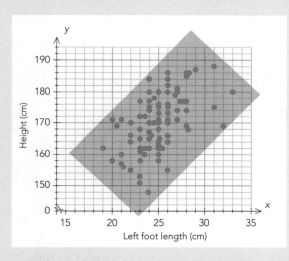

I notice that there is **moderate** amount of scatter between the points on the graph.

This means there is a **moderate** relationship between left foot length and height but it **isn't strong**.

2 The relationship between students' grades and the amount of time in the evening spent on social media.

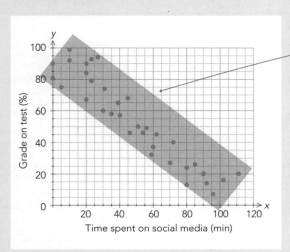

Notice the shading is narrower than in the first example.

I notice that there **isn't much scatter** between the points on the graph.

This means there is a **strong** relationship between the time spent on devices and the grade in a test.

Complete/write the statements about the strength of the relationships below.

4   This graph shows the relationship between age (years) and net worth ($millions) of some celebrities.

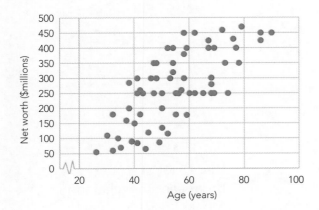

I notice that there _____

_____ amount of

scatter between the points on the graph.

This means _____

_____

_____

5   This graph shows the relationship between lowest daily recorded temperature and highest daily recorded temperature in Arthur's Pass.

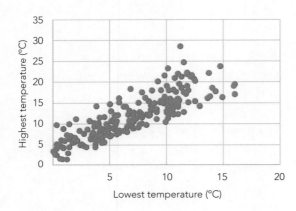

I notice that there _____

_____ scatter

between the points on the graph.

This means _____

_____

_____

6   This graph shows the relationship between distance travelled (km) to school and the time taken (min).

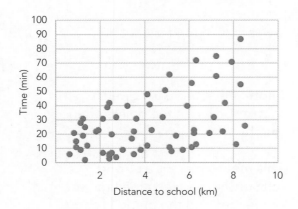

_____

_____

_____

_____

_____

_____

_____

_____

_____

_____

Trend

Linear

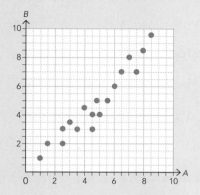

A line drawn through the middle of the data **would** be straight.
This means the data **is** changing at a constant rate.

Non-linear

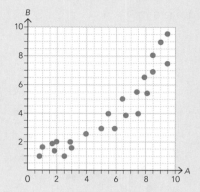

A line drawn through the middle of the data **would not** be straight (curved).
This means the data is **not** changing at a constant rate.

Examples:

1 The graph shows the relationship between the lengths of students' left feet and their heights.

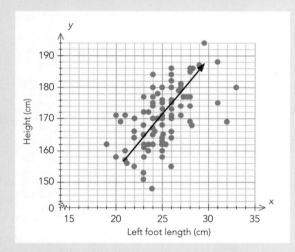

I notice the **trend** appears to be linear **because** I could rule a **straight line** through the middle of the data.

This means that as the length of student's left foot increases, the height tends to increase at **a constant rate**.

2 The graph shows the relationship between the ages of a group of children and the time they took to complete a puzzle.

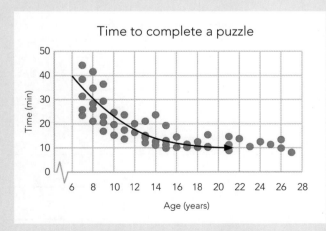

I notice the **trend** appears to be non-linear **because** a line drawn through the middle of the data would be a curve.

This means that as the child's age increased, the time they took to complete a puzzle does not decrease at a constant rate.

Complete/write the statements about whether these relationships are linear or not.

7 This graph shows the relationship between the sizes of districts and the number of public toilets in each.

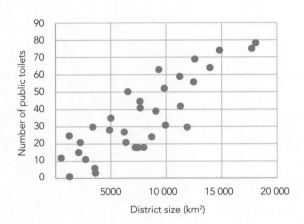

I notice the **trend** appears to be linear/non-linear **because** I could/couldn't rule a straight line through the middle of the data.

This means that as the district size increases, the number of public toilets does/doesn't tend to increase at a constant rate.

8 This graph shows the relationship between the number of kilometres that a car has driven and its price.

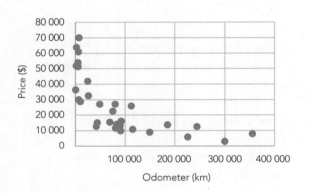

I notice the **trend** appears to be

_____ **because** I

_____ rule a straight line through the middle of the data.

This means that as the _____

increases, the _____

_____ tend to decrease at a

constant rate.

9 This graph shows the relationship between the strength of the average wind gusts and the strengths of the maximum gusts at 800 m in Arthur's Pass National Park.

_____

_____

_____

_____

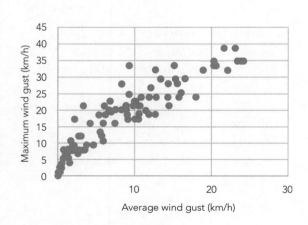

_____

_____

_____

_____

Putting it all together

Example: The relationship between students' ages and their heights.

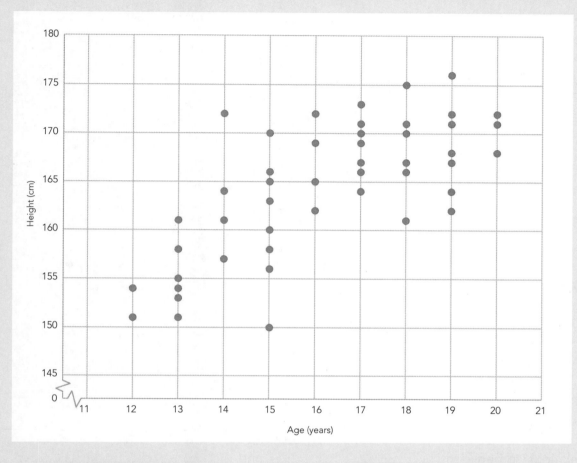

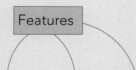

There is a **positive**, **moderate**, **linear** relationship between age and height.

Direction: I notice that the relationship between a student's age and their heights is positive. This means as students grow older, their height tends to increase.

Strength: I notice that there is moderate amount of scatter between the points on the graph. This means there is a relationship between age and height but it isn't strong.

Trend: I notice the trend appears to be linear because I could rule a straight line through the middle of the data. This means that the relationship between age and height changes at a constant rate.

Put it together for the following sets of data.

1 The graph shows the relationship between the number of pies a bakery sold each day and the maximum daily temperature.

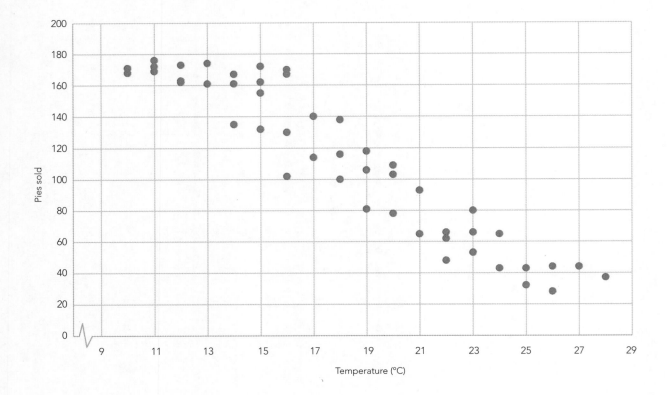

There is a _____, _____, _____

relationship between _____ and _____.

Direction: I notice that the relationship between _____ and

_____ is _____.

This means as the _____ increases, the

_____ tends to _____.

Strength: I notice that there _____ scatter between the

points on the graph. This means _____

_____

Trend: I notice the trend appears to be _____ because

_____

This means _____

_____

2   This graph shows the relationship between the number of students and number of staff in New Zealand schools.

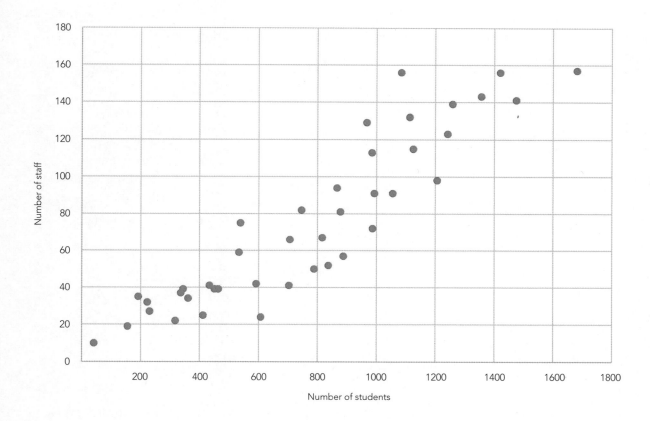

There is a _____

_____

_____

Direction: _____

_____

_____

Strength: _____

_____

_____

Trend: _____

_____

_____

Unusual features

Unusual points

- These are points which lie some distance from most of the other points.
- Sometimes these are valid points.
- Otherwise they may be a result of measurement error, recording error, or reversed coordinates.

Example:

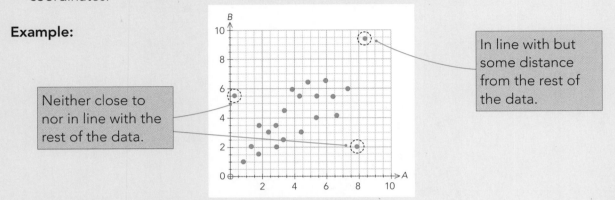

Neither close to nor in line with the rest of the data.

In line with but some distance from the rest of the data.

Unusual clusters of points

- These are small groups which lie some distance from the rest of the data.
- You should always identify these, and try to find an explanation.

How many make a cluster?
Don't get hung up on labels and definitions. It is best to write what you see. If there are two points away from the rest, then say that.

Examples:

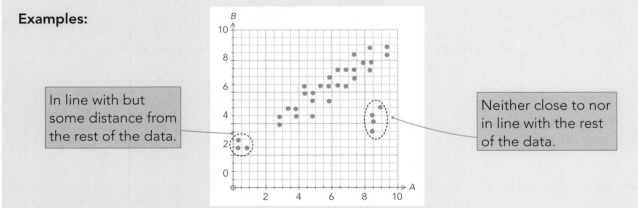

In line with but some distance from the rest of the data.

Neither close to nor in line with the rest of the data.

Groupings may overlap and not become evident until you use different colours for different groups within your data, e.g. males and females.

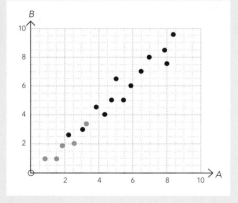

Unusual points and clusters must be *very* obvious. Not every data set has unusual points or clusters.

 ISBN: 9780170477437

Example: Consider the arm spans and heights of a group of Year 11 students.

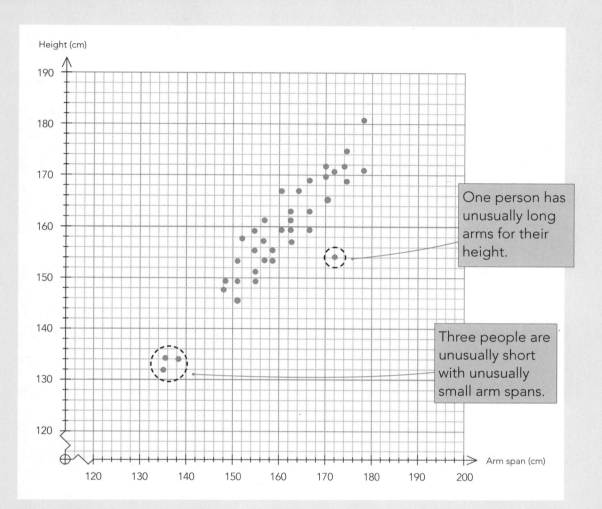

What do you see?

I notice that there is a cluster of smaller values and one larger value that are significant distances from most of the data.

Why might we have these clusters or unusual features? Link to the context if possible.

What does this mean for the sample?

This means that there is one student who, given their arm span (172 cm), is unexpectedly short (154 cm) compared with the rest of the students. This student's arm span and height are not in the same proportion as the rest of the students because their point is a long way from the rest of the data.

There are also three students who are both much shorter than most (around 133 cm), and have smaller arm spans (around 137 cm). Although these students are much smaller than the others, their arm span is in proportion to their height because their points are in line with the rest of the data. A possible explanation is that these students are much younger than the other students.

Answer the following questions.

1 This graph shows the relationship between the time spent studying and the grade in a test.

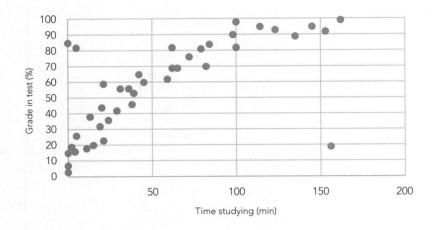

I notice that _____

_____

This means that _____

_____

2 This graph shows the relationship between the age of famous artworks and their estimated value.

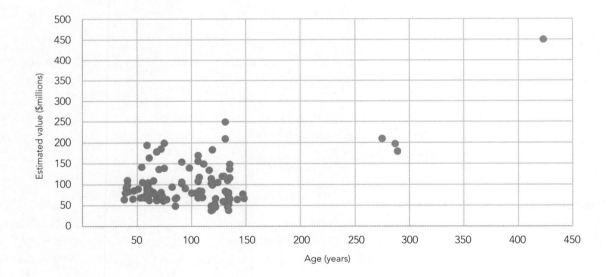

I notice that _____

_____

This means that _____

_____

Variation in scatter

- Sometimes the variation in scatter changes across the graph.
- This means that the relationship is stronger in some parts than others.

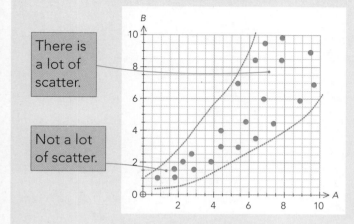

There is a lot of scatter.

Not a lot of scatter.

I notice that the scatter in this graph isn't very even.
The data for the smaller values of the graph are closer together than the data for the larger values on the graph.

This means that there is a stronger relationship between A and B in the smaller values of the graph and a weaker relationship between A and B with the larger values on the graph.

Describe the variation in the scatter in these graphs and what this means.

 11

1 The graph shows the ages of New Zealand football players and number of games they have played.

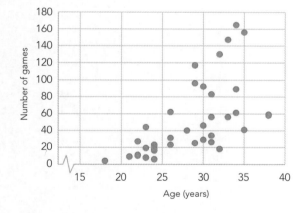

I notice that _____

_____

_____

This means that _____

_____

_____

_____

2 This graph shows the daily rainfall (mm) and number of visitors to a zoo.

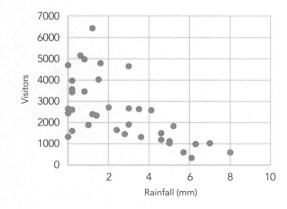

I notice that _____

_____

_____

This means that _____

_____

_____

_____

Line of best fit

- It is useful to add a line of best fit to a scatter plot. Most computer programs will do this for you.
- The line of best fit:
  — should be **ruled**
  — should have about the same number of points above and below it
  — should pass through the middle of the points at both ends.

Examples:
Good line of best fit:

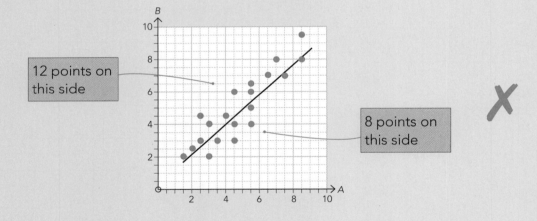

Passes through the middle of data at each end.

8 points on each side of the line

✓

Poor lines of best fit:

12 points on this side

8 points on this side

8 points on both sides, but not through the middle at both ends.

✗

Hint: It really helps to have a **clear** ruler for this if you are drawing by hand.

Use a ruler to draw a line of best fit on the following graphs.

1

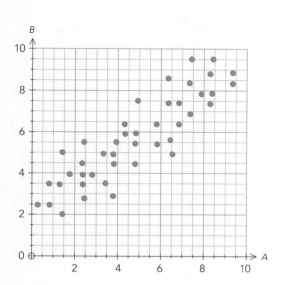

2

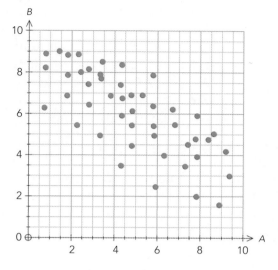

3

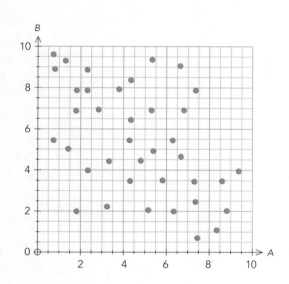

4

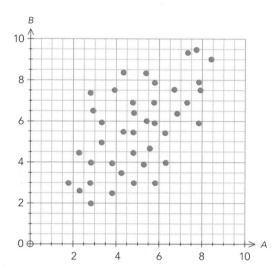

5

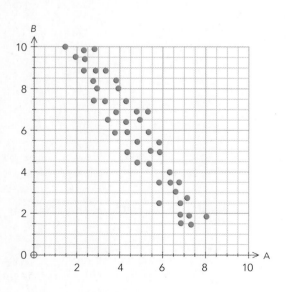

6

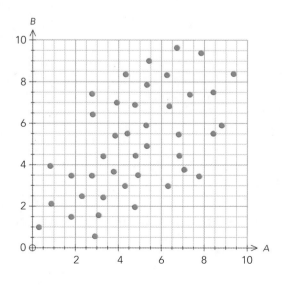

Predictions

- A line of best fit can be used to confirm features and to help to estimate predictions.
- Because we are using mathematical models to describe real situations, we can never be certain that predictions will be accurate.
- How precise the prediction is will depend on:
  — the strength of the relationship
  — the amount of data
  — whether the data is a true representation of the population.

Example: Make a prediction for the height of a student whose foot length is 29 cm.

Step 1: Draw a line of best fit. ———
Step 2: Estimate the point value. - - - - ➤

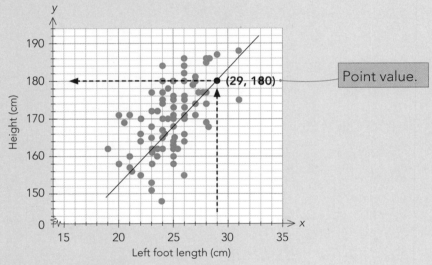

Step 3: Shade the variation in the data.
Step 4: Use the boundaries of your shading to estimate the range within which the height could fall.

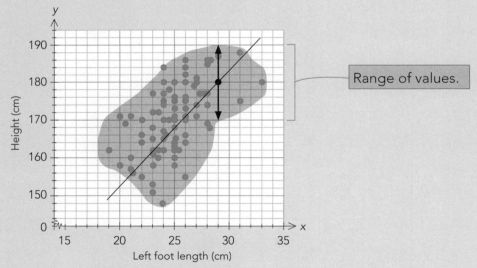

Estimate:
I predict that someone with a foot length of 29 cm **will be about** 180 cm tall, **but** their height **could be anywhere between** 170 cm and 190 cm.

Confidence:
I **am reasonably** confident in my prediction because the strength of the relationship is moderate.

1 The graph shows the relationship between the number of pies sold by a bakery each day and the maximum daily temperature.

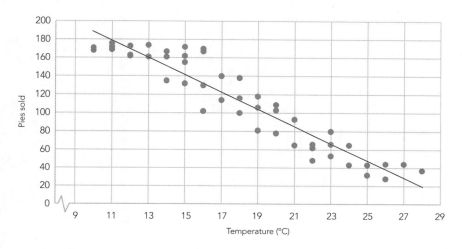

a I predict that if the temperature is 23°C, the number of pies sold will be about _____

but could be anywhere between _____ and _____.

b I predict that if the temperature is 16°C, the number of pies sold will be about _____

but could be anywhere between _____ and _____.

c I am very/reasonably/not very confident in my predictions because the relationship is strong/moderate/weak.

2 The graph shows the relationship between the number of students on a New Zealand secondary school roll and the number of staff.

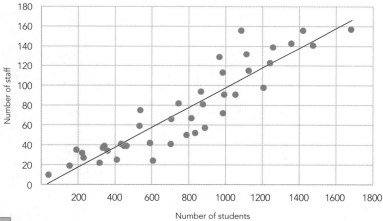

Choose your own value

a I predict that a school with _____ students will have about _____ staff but could

be anywhere between _____ and _____.

b I predict that a school with _____ students will have about _____ staff but could

be anywhere between _____ and _____.

c I am _____ confident in my predictions because the relationship is _____.

3 This graph shows the relationship between age (years) and net worth ($millions) of some celebrities.

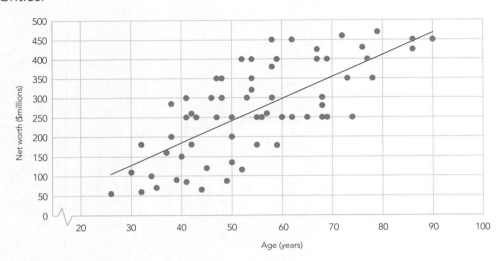

a I predict _____

but _____.

b I predict _____

but _____.

c I am _____.

4 This graph shows the relationship between age and height of students.

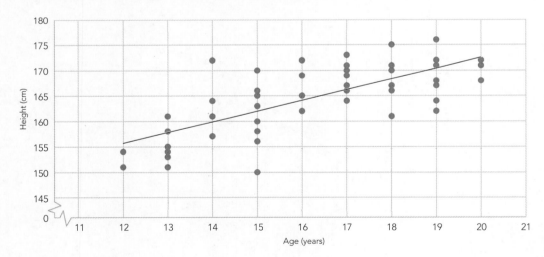

a I predict _____

_____.

b I predict _____

_____.

c I am _____.

Putting it all together

You will need to do the following:

1 Introduction
- Briefly **describe** the context of your investigation.
- Write or identify the investigative **question** or statement.
- Describe a **purpose** for your investigation.

2 Investigation details

Collecting your own data	Given data
a Identify the **variables** you are collecting. **b** Consider and describe where your sample data is going to come from. **c** Describe in detail how you are **selecting** or **measuring** your data. **d** State how many pieces of data you collect.	**a** Identify the **variables** you have been given. **b** Describe where your sample data has come from. **c** State how many pieces of data you have.

3 Variation

Collecting your own data	Given data
Describe at least two different types of **variation**, and how you will manage these.	Describe at least two different types of **variation** that may have occurred, and your **assumptions** about how they have been managed.

4 Display your results
- Create a **scatter plot** to show your results.

5 Describe what you see in your scatter plot
- Describe the **direction** of the scatter and **explain** what it means.
- Describe the **strength** of the scatter and **explain** what it means.
- Describe the **trend** of the scatter and **explain** what it means.
- Describe **other features** such as inconsistent scatter and unusual points or groupings.

6 Add a line of best fit

7 Make a prediction
- Comment on your confidence.

8 Write a conclusion
- **Answer the question**
  - Do your results seem **reasonable**?
  - Who does your conclusion apply/not apply to?
  - Would you expect to get the same results if you repeated your investigation?
- **Reflections**
  - How could you have improved the reliability of your results?
  - Are there other factors that you haven't mentioned so far that might have affected your results?
  - What are the next steps? What could this investigation lead to?

walkermaths

Annotated example

The blue duck, or whio, is a species that is found only in New Zealand. Their numbers are estimated to be fewer than 3000.

https://www.doc.govt.nz/nature/native-animals/birds/birds-a-z/blue-duck-whio/

1 Introduction

I wonder if there is a relationship between the mass of whio and the wing length for whio in the Tasman area. Probably, heavier birds will need larger wings to enable them to fly effectively. If we could use the weight of the birds to estimate their wingspan, it may reduce the need to handle them and therefore not stress them more than necessary.

Question

Purpose

Scientists collect anatomical data on this species from the Tasman region. The two variables we are investigating are the body mass in grams and length of one wing (mm) of whio from the Tasman area. We have a sample of 138 whio.

2 Investigation
   details

Given the data was collected for us, we assume that variation was managed appropriately by the researchers. Here are some potential sources of variation:

3 Variation

- Measurement variation because the ducks were unlikely to have stayed still. A common technique to keep birds calm is to cover their heads with a breathable cloth. We assume the researchers were trained in appropriate bird-handling techniques to reduce stress and increase accuracy of measurement.
- Situation variation because we don't know how recently the birds have eaten. This could be managed by measuring birds at similar times of the day.

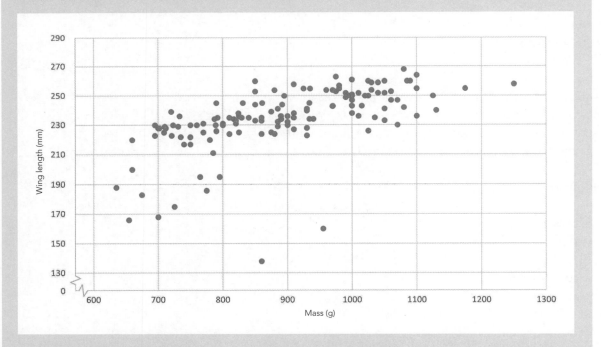

4 Display

6 Line of
   best fit

I notice there is a moderate, positive, linear relationship between whio mass and wing length.

5 Description

I notice that the relationship between mass (g) and wing length (mm) in whio is positive. This means as the mass of the bird increases, the wing length tends to increase.

Direction

ISBN: 9780170477437

I notice that there is a moderate amount of scatter between the points on the graph. This means that there is a moderate relationship between mass (g) and wing length (mm) in whio but it is not strong.

Strength

I notice that the trend appears to be linear because I could rule a straight line through the middle of the data. This means that as the mass (g) increases, the wing length (mm) in whio tends to increase at a constant rate.

Trend

There are a few birds that are unusual. Two of them have a bigger mass than we would expect for their wing length. One bird has mass 860 g and the smallest wing length (138 mm). The other has mass 955 g and a wing length of 160 mm. There is also a group of whio that have smaller masses and shorter wings than expected. From research young whio spend most of their time swimming in rapids, so it may be that their wings do not develop to their full length until they are older and heavier.

Other features

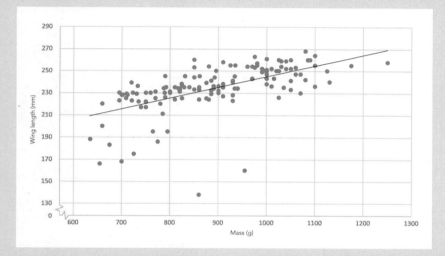

If a whio's mass was 1100 g, we would expect their wing length to be around 250 mm. However, it could be anywhere between 235 and 270 mm. I am reasonably confident in my prediction because the strength of the relationship is moderate.

7   Prediction

Conclusion

There is a moderate, positive, linear relationship between mass and wing length of whio from the Tasman region. This means that the larger the mass of the whio, the longer their wing length is. This result could apply to other whio in the Tasman area but may not be applicable to whio in other areas or other species of ducks.

8   Conclusion

If another sample was taken, we would likely be measuring different whio and would therefore get different data. Given the sample size was sufficient, it is likely that we would see a similar trend to the one found above. To be able to apply our findings to whio in the Tasman area, we are assuming that the sample is representative of this population.

Prediction and Conclusion

Reflections
- If a larger sample of whio were taken from the Tasman area, there would be more data and it would likely make the relationship clearer. We would be more confident in our conclusions.
- Next steps could be to look into whether there are differences between juvenile and adults or male and female whio. You could also look at whio from other parts of New Zealand.

Reflections

Practice task

Introduction
Orange-spotted geckos are one of the 48 species of geckos found in New Zealand. They tend to be found in mountainous areas and are largely active at night.
https://www.doc.govt.nz/nature/native-animals/reptiles-and-frogs/lizards/geckos/orange-spotted-gecko/

Investigation details
This data was collected by researchers in Otago.
SVL (snout–vent length) is the measurement (in millimetres) on a reptile from snout to vent.
Mass is the other variable, measured in grams. We have a sample of 58 gecko.

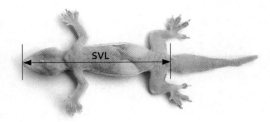

Question and purpose

_____

_____

_____

Variation

_____

_____

_____

_____

_____

Graph

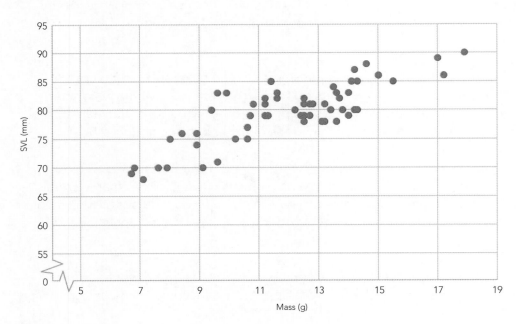

Description

_____

_____

Direction

_____

_____

_____

Strength

_____

_____

_____

Trend

_____

_____

_____

Other features

_____

_____

_____

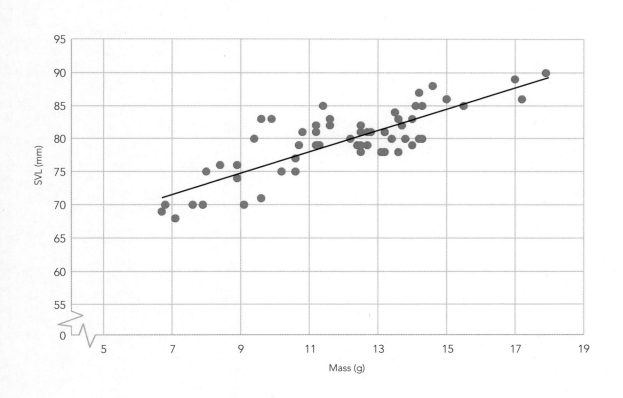

Prediction

_____

_____

_____

_____

_____

Conclusions

_____

_____

_____

_____

_____

_____

_____

Reflections

_____

_____

_____

_____

_____

_____

_____

_____

Comparison

- **Comparing** data can enable us to make judgements and come to conclusions about **differences between groups**.
- Graphs that can be used for this include dot plots, box plots or back-to-back bar graphs.
- Which graph you choose will depend on the type of data that you have.

For **comparative** investigations, we use a sample to make an **inference** (suggestion) about the two (or more) populations.

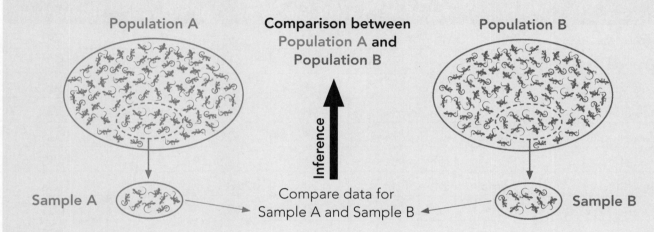

What's the point?

- The point is to **compare two (or more) different groups** using **numerical data**.
- The purpose of your statistical investigation should be clear and defined.
- Posing a question can help to clarify this.

Example: I wonder if Year 13 students **tend to be** taller than Year 11 students at Paradise High School.

> Avoid being definite because you can never be certain that every Year 13 student will be taller than every Year 11 student.

What do we need to think about?

What are the populations? What populations is your inference about?

Type of data — you must have **numerical** data for two or more groups.

> This can be discrete or rounded continuous data.

Sample size — each group should have **at least 30** pieces of data for each group.

Where am I getting my data?

If you are given it, where did it come from?

If you are collecting it, you must describe how you did this.

> The larger the sample, the more likely it is to be representative of the population.

Variation

Explain at least two different types of variation and how these could be managed.

Variable selection

- When undertaking a comparison investigation, **one numeric variable** and **one categorical variable (with two groups)** should be selected.
- It's sensible to think about how or why one group maybe different to the other. This can help lead to interesting discussion.

There are several species of gecko in New Zealand. The table below shows data for two species. Elevation is the height above sea level that the gecko was found. SVL is the measurement of a reptile from snout to vent.

Species	Sex	Pattern	SVL (mm)	Mass (g)	Age	Elevation (m)
Orange spotted	Male	no orange	80	14	Adult	1510
Orange spotted	Female	orange spots	69	14.5	Juvenile	1519
Orange spotted	Male	spots	80	12.4	Adult	1500
Jewelled	Male	diamonds	74	11.2	Juvenile	1508
Jewelled	Male	barbed wire	75	10.6	Adult	1506
Jewelled	Female	diamonds	78	13.8	Adult	1488
Jewelled	Female	diamonds	71	10.1	Juvenile	1423
Orange spotted	Male	no orange	81	13.2	Adult	1404
Orange spotted	Male	orange spots	80	11.5	Adult	1377
Jewelled	Female	diamonds	67	10.9	Juvenile	1579
Orange spotted	Female	no orange	61	12.3	Juvenile	1581
Jewelled	Male	barbed wire	78	12.1	Adult	1583
Orange spotted	Male	orange spots	77	10.9	Adult	1583
Orange spotted	Female	orange spots	70	11.8	Adult	1582
Orange spotted	Female	orange spots	73	10.5	Adult	1512
Jewelled	Female	diamonds	76	11.9	Adult	1500

What variables from this data set would be appropriate to be used in a comparison investigation?

Numeric variable	Groups	Reason for investigating these variables
SVL	Male and Female	SVL of male gecko may be different to the SVL of females because females bodies need to be able to produce and carry eggs.

Dot plots

- A dot plot provides a visual representation of the data.
- Dot plots are used for **discrete** and **rounded continuous** data.
- Each dot represents one person/object, unless you are told otherwise.
- Dot plots are particularly useful because they show the value of every piece of data.

Understanding dot plots

Examples:

1 A **single** dot plot answer can be used to answer a **summative** question.
Students at Paradise High School were each asked to record how many plane flights they had taken this year.

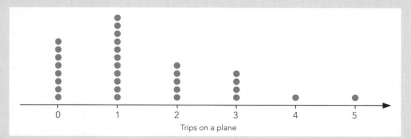

Possible question: What was the most common number of plane flights taken by Paradise High School students this year?

2 A **pair** of dot plots can be used to answer a **comparative** question.
Year 11 and Year 13 students at Paradise High School were asked how long they spent doing homework the previous night.

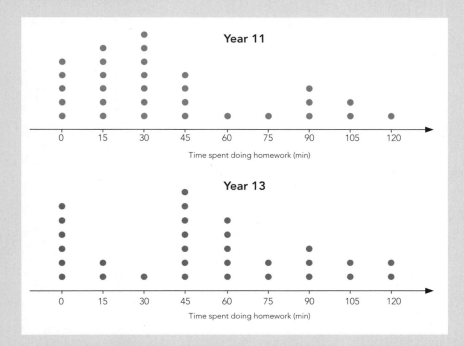

Possible question: At Paradise High School, do Year 13 students tend to spend longer doing homework than Year 11 students?

During this block of work, we will be dealing with comparative questions.

Answer the following questions. Round any calculations to 3 sf.

1 Question: 'How many movies did students see in the last year?'

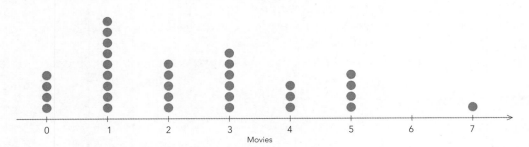

a What type of question is this?   Summative/Comparative

b How many students saw exactly three movies?   _____

c How many students saw fewer than two movies?   _____

d What was the most common number of movies seen?   _____

2 These graphs show the ages of the All Blacks and Black Ferns.

All Blacks

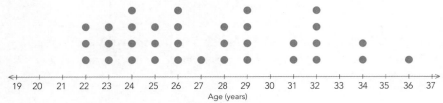

Black Ferns

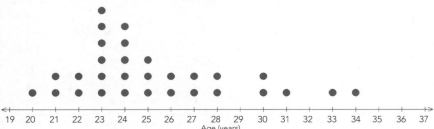

a How many All Blacks are over 30 years of age?   _____

b What is the most common age of the Black Ferns?   _____

c Which team tends to be made up of older players?   _____

How do you know? _____

d Most All Blacks are between 20 and 30 years old.

☐ Agree   ☐ Disagree   ☐ Can't tell for sure

Explain your answer. _____

_____

Describing features of dot plots

From a dot plot you can identify the centre, the spread, unusual features and the shape.

Centre
This can be quantified by calculating the median, mean or mode. The median and mode are easiest to identify from a dot plot.

Spread
This can be quantified by calculating the range.

Unusual features
- These could be unusual points or clusters.
- An unusual point is one that is a long way from the rest of the data.
- A cluster is a group of data that is away from the rest of the data.

How many make a cluster?
Don't get hung up on labels and definitions. It's best to write what you see. If there are two points away from the rest, then say that.

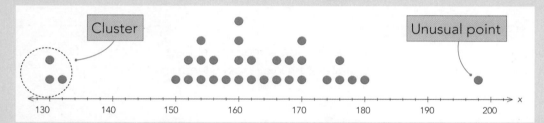

Note:
1 Check that an unusual point is not a mistake in measurement, counting or recording.
2 Include unusual features when you graph and analyse your data. Think about whether the feature is possible and state your thoughts when you discuss the data.

1

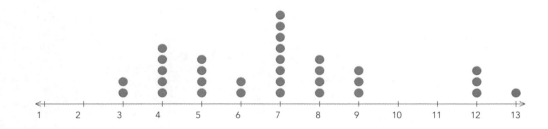

    a   What is the mode?   _____   **b**   What is the median?   _____

    c   Calculate the range. _____

    d   Identify and describe any unusual features.

_____

_____

2 Students were asked 'At what age do you think young people should be allowed to drive in New Zealand?'

Age at which young people should be allowed to drive in New Zealand

a What is the mode? _____ **b** What is the median? _____

c Calculate the range. _____

d Identify and describe any unusual features.

_____

_____

3 A class compared two brands of hokey-pokey ice cream. Each student ate two scoops of each brand and recorded how many pieces of hokey-pokey were in each.

Polar Ice

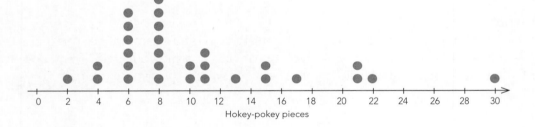

Hokey-pokey pieces

Nice Cream

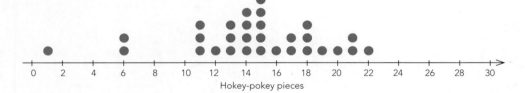

Hokey-pokey pieces

a Compare the modes. _____

b Estimate and compare the medians. _____

c Compare the spreads. _____

d Identify and describe any unusual features.

_____

_____

 ISBN: 9780170477437

Shape

If samples are large enough, you may be able to comment on the shape.
None will be perfect, so use the term '**tends towards**'.

Bell shaped/normal distribution Symmetrical hill or mound shape.	
Right skew The tail is longer on the right-hand side.	
Left skew The tail is longer on the left-hand side.	
Bimodal There are two peaks.	
Uniform Looks like a box.	
Irregular There is no real pattern.	

- Some computer programs allow you to add a shape outline to a dot plot. This can make the distribution easier to visualise.
- This can also be done by hand.

Example:

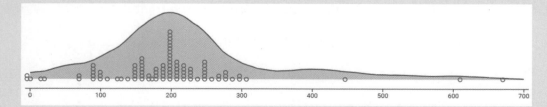

This graph tends towards a right skew. Most of the data is to the left, with a longer tail to the right.

ISBN: 9780170477437

Highlight the shape that these graphs tend towards.

4

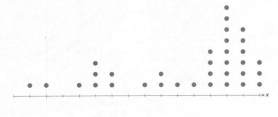

Bell shaped   Irregular
Right skew   Bimodal
Left skew   Uniform

5

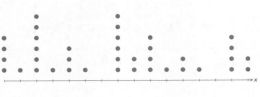

Bell shaped   Irregular
Right skew   Bimodal
Left skew   Uniform

6

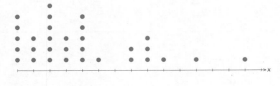

Bell shaped   Irregular
Right skew   Bimodal
Left skew   Uniform

7

Bell shaped   Irregular
Right skew   Bimodal
Left skew   Uniform

8

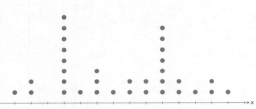

Bell shaped   Irregular
Right skew   Bimodal
Left skew   Uniform

9   Compare these distributions.

A

B

Centre: The centre of distribution _____ is further to the right than that of distribution _____.

Spread: Distribution _____ is more spread out than distribution _____.

Shape: Distribution A tends towards _____

and distribution B tends towards _____.

Comparing dot plots

- Here are 30 pieces of data for two groups. Compare centre, spread, shape and unusual features.

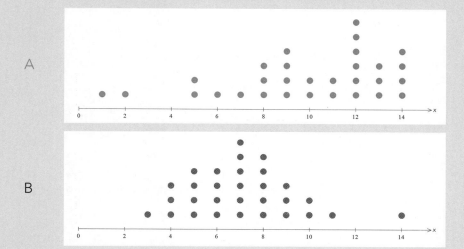

Centre: The centre of distribution A (10 or 11) is further to the right than that of distribution B (6 or 7).

> Always include **numbers** to support centre and spread statements.

Range: The range of A (13) is larger than the range of B (11).

Shape: The shape of A tends towards a left skew, whereas the shape of B tends towards a bell-shaped distribution.

Other features: B has one unusual point (14), which is some distance from the rest of the data. A has two values (1 and 2) that are smaller than the rest of the data.

Indicate whether these statements are correct ✓ or incorrect ✗.

1

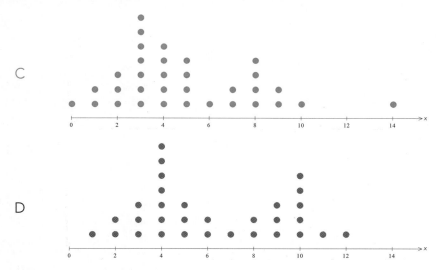

a The range of C is larger than the range of D. ☐

b The shape of C tends towards a left skew and D tends towards a bimodal distribution. ☐

c Neither has any unusual features. ☐

d The mode of D is larger than the mode of C. ☐

2 A group of students got their results back from two recent tests.

Science

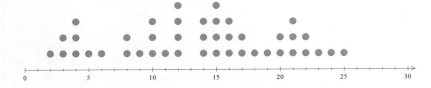

English

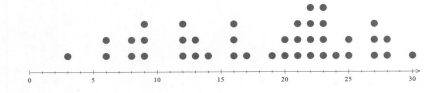

a The Science scores are more variable than the English scores. ☐

b Both data sets tend towards a bell shape. ☐

c The highest score was in English. ☐

d There are no unusual features in either data set. ☐

3 The data shows the number of eggs laid per day in two chicken coops. Compare the features of these two graphs.

Pina's coop

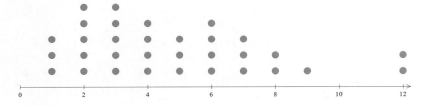

Gary's coop

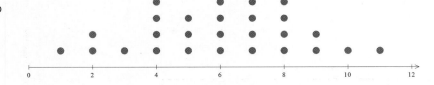

_____

_____

_____

_____

Describing meanings of features

Tomato seedlings

Huia grows tomato plants to sell at the market. She sows one
seed into each of six pots in one tray. She grows seeds of two
different types of tomato: Tasty Tommy and Mighty Monster.
She sows 30 trays of each type of tomato and records the
number that germinate in each tray.

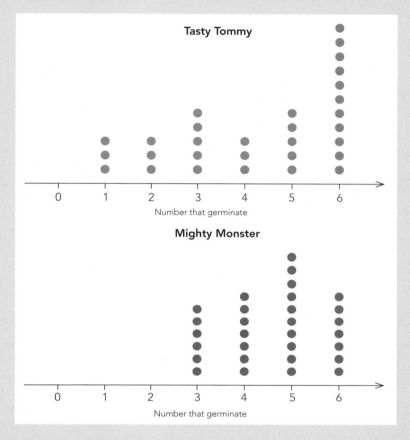

Centre

I notice that the median of germinated seeds per tray for Tasty Tommy (5) is the same as the
median of germinated seeds per tray for Mighty Monster (5).

This means that both tomato types of seed have similar germination rates.

Spread

I notice that the range for Tasty Tommy (5) is larger than the range for Mighty Monster (3).

This means that the numbers of Tasty Tommy seeds that germinate in each tray are more
widely varied than those for Mighty Monster.

Shape

I notice that the distribution for the number of germinating seeds per tray for Tasty Tommy
seeds tends towards a left skew and the data for Mighty Monster seeds tends towards a normal
distribution.

This means that more trays of Tasty Tommy seeds have only one or two germinate, but every
tray of Mighty Monster seeds had at least three germinate.

Indicate whether these statements are correct or incorrect.

1

C
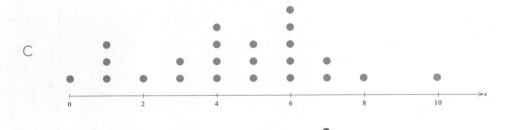

D
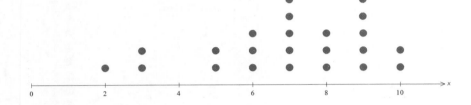

Centre: I notice that the median of set C (_____) is **smaller/larger** than the median of

set D (_____).

This means:

☐ The sample size of set C is smaller than set D.

☐ The data is more spread in data set D than in data set C.

☐ The items in set C tend to be smaller than those in set D.

☐ The items in set D are all larger than those in set C.

Spread: I notice that the range of set C (_____) is **smaller/larger** than the range of

set D (_____).

This means:

☐ The sample size for D wasn't large enough.

☐ There is more variation in data set C than there is in data set D.

☐ There is more variation in data set D than there is in data set C.

Shape: I notice that data set C tends towards a **bell/bimodal/uniform/irregular** shape.

This means:

☐ If we had more data, we might see a pattern for set C.

☐ There is no particular pattern to the distribution.

I notice that data set D tends towards a **left/right** skew.

This means:

☐ The data is more spread out on the left hand side.

☐ The data in set D is useful and the data in set C is not.

2   Two friends compare the amount of honey their bees produced.

Bernie's bees

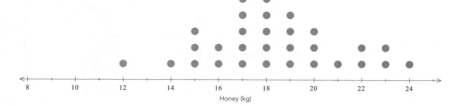

Hemi's hives

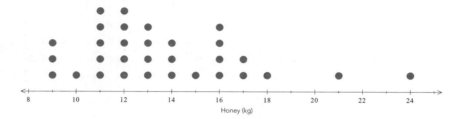

Centre: I notice that the median amount of honey of Bernie's bees (_____) is **smaller/larger**

than the median amount of honey from Hemi's hives (_____).

This means:

☐ Hemi has more bees than Bernie.

☐ Bernie's bees tend to produce more honey than Hemi's.

☐ Bernie's bees are bigger than Hemi's.

☐ All Bernie's bees produce more honey than Hemi's.

Spread: I notice that the range of honey produced by Bernie's bees (_____) is **smaller/larger**

than the range of honey produced by Hemi's hives (_____).

This means:

☐ Bernie's bees are better than Hemi's.

☐ There is more variation in the amount of honey produced by Bernie's bees than in Hemi's hives.

Shape: I notice that the distribution for Bernie's bees tends towards a **bell/bimodal/uniform/ irregular** shape.

This means:

☐ No more data should be collected.

☐ The hives are producing their maximum amount of honey.

☐ The data is symmetrically distributed with most of the data in the middle.

I notice that Hemi's hives tends towards a **left/right** skew.

This means:

☐ Most of the data is on the right.

☐ The data is more spread out on the right-hand side.

3 During a week-long fishing competition, a crew keeps track of the length of their catch.

Rāwaru (blue cod)

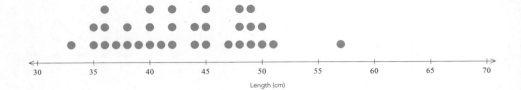

Mararī (butterfish)

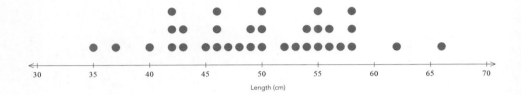

Centre: I notice _____

_____

_____

This means _____

_____

_____

Spread: I notice _____

_____

_____

This means _____

_____

_____

Shape: I notice _____

_____

_____

This means _____

_____

_____

Box plots

- These are used to display discrete or rounded continuous data.
- They are sometimes called 'box and whisker plots'.
- These are very useful for comparing sets of data.

For a box plot, you need to know the following:

Minimum   **Lower quartile**   **Median**   **Upper quartile**   **Maximum**

Example:

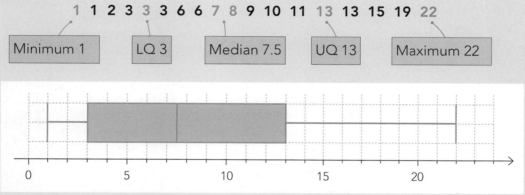

1 **1 2 3** 3 **3 6** 6 **7** 8 **9 10 11** 13 **13 15 19** 22

| Minimum 1 | LQ 3 | Median 7.5 | UQ 13 | Maximum 22 |

Other useful statistics

range = maximum − minimum

interquartile range = upper quartile − lower quartile (often shortened to IQR)

Answer the following questions.

1   Use the words in the box to complete this diagram.

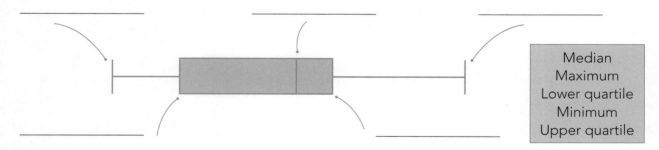

| Median |
| Maximum |
| Lower quartile |
| Minimum |
| Upper quartile |

2   Draw a box plot with these statistics:

Minimum = 2   LQ = 5   Median = 12   UQ = 16   Maximum = 19

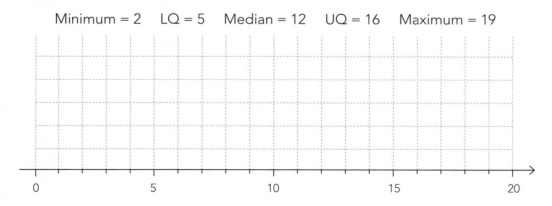

ISBN: 9780170477437

Understanding box plots

You need to understand the meanings different parts of a box plot.

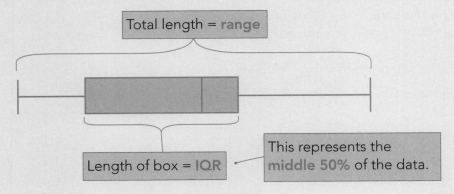

Data is symmetrically distributed ⇒ box plot will be symmetrical.

Data is not symmetrically distributed ⇒ box plot will not be symmetrical.

Example:

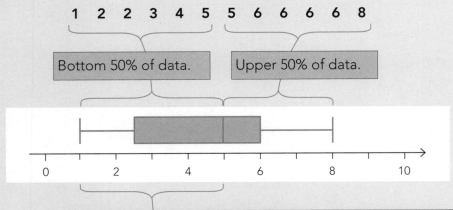

The **vertical lines** on a box plot divide the data into **quarters**.

         ISBN: 9780170477437

Fill in the missing percentages and terms below.

1

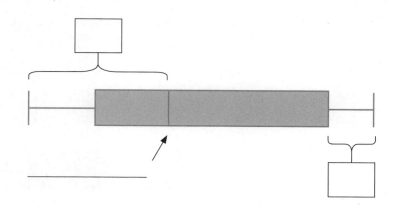

2

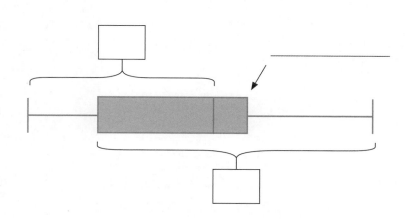

3

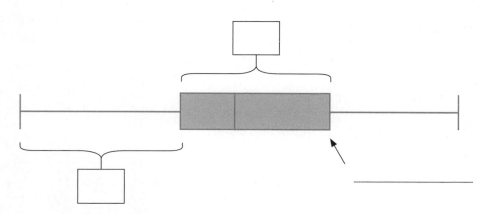

4

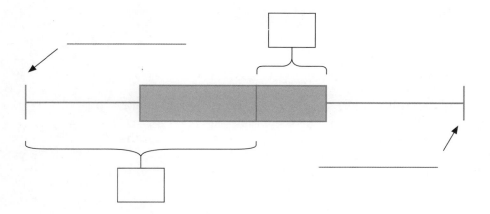

Comparing box plots

- Box plots are a useful way of comparing several groups of data.

Example:

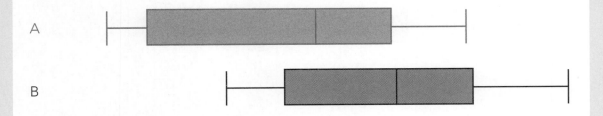

The median of A is higher than the median of B. ✗
The interquartile range of A is larger than that of B. ✓
The upper quartile of A is lower than the median of B. ✓
The range of A is smaller than the range of B. ✗

Identify if these statements are correct or incorrect.

1

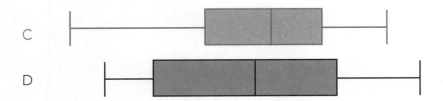

a   The median of C is larger than the median of D. ☐

b   The upper quartile of D is lower than the median of C. ☐

c   Half of the data in C is larger than the median of D. ☐

d   The IQR of D is larger than the IQR of C. ☐

2

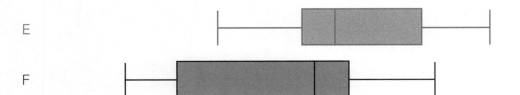

a   The median of E is larger than the upper quartile of F. ☐

b   The lower quartile of F is smaller than the minimum of E. ☐

c   Three quarters of the data in F is smaller than half of the data in E. ☐

d   Half of the data in E is larger than half the data in F. ☐

3

Height of students (cm)

Year 9

Year 12

a 75% of the Year 12 students were taller than 50% of the Year 9 students. ☐

b Half of the Year 9 students were taller than the LQ of Year 12 students. ☐

c The tallest Year 9 student was the same height as the UQ of Year 12 students. ☐

d The shortest Year 12 student was the same as the LQ of Year 9 students. ☐

4

Length of toilet paper rolls (m)

Softex

Boggy

a The median length of Boggy rolls was longer than the median length of Softex. ☐

b The upper quartiles of both brands were similar. ☐

c Boggy had more variability in roll length than Softex. ☐

d Softex had the longest roll. ☐

5

Reaction time (s)

Right handed

Left handed

a The student who took the longest time was right handed. ☐

b The quickest reaction times of right handed and left handed were similar. ☐

c Half of left-handed students were quicker than three quarters of right handed. ☐

d The median right-hand reaction time was faster than the median of left-hand time. ☐

Describing meanings of features

Wiremu has a bird feeder to attract more tūī and korimako (bellbirds) to his garden. He recorded the length of each visit to the feeder. He constructed box plots to show the differences between the two species.

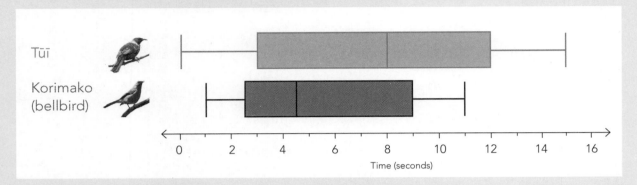

Centre

I notice that the median time spent feeding by tūī (8 s) is further to the right than the median feeding time for korimako (4.5 s).

This means that the median time that tūī spent feeding is longer than that of the korimako.

Spread

I notice that the interquartile range of tūī (9 s) is wider than the IQR for korimako (6.5 s).

This means that there is more variation in the lengths of feeding times for tūī than korimako.

Shape/symmetry

Tūī
I notice that the four sections of the box plot for tūī are similar lengths.

This means that the distribution of times for tūī is reasonably symmetrical and the times are evenly spread.

Korimako
I notice that intervals in the two left sections of the box plot for korimako are shorter than those on the right, so the data is skewed to the right.

This means that there is less variation in the lengths of short visits by korimako, and the lengths of longer visits are more variable.

16

1   Henare's mum owns an ice-cream cart. The most popular flavours are chocolate and
    coconut. Henare recorded the numbers of each of these flavours that were sold each day.
    His results are shown in these box plots.

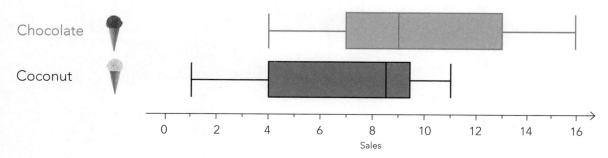

a   Centre

I notice that the median number of chocolate ice creams sold (9) is higher than the
median number of coconut ice creams sold (8.5).

This means _____

_____

_____

b   Spread

I notice that the IQR for the number of chocolate ice creams sold (6) is similar to that for
coconut (5.5).

This means _____

_____

_____

c   Shape

Chocolate

I notice _____

_____

This means _____

_____

_____

Coconut

I notice _____

_____

This means _____

_____

2 Harriet collected hourly pay rates for students working in construction and those working in IT. She created box plots to show the data.

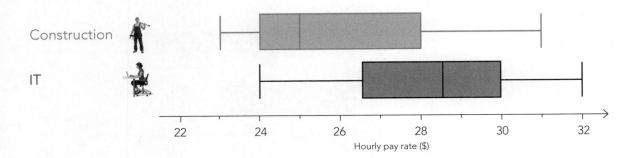

Construction

IT

Hourly pay rate ($)

a **Centre**

I notice _____

This means _____

_____

_____

b **Spread**

I notice _____

This means _____

_____

_____

c **Shape**

Construction

I notice _____

_____

This means _____

_____

_____

IT

I notice _____

_____

This means _____

_____

Making the call

- We can make definitive statements about the **sample** by comparing the box plots.
- However, the purpose of comparing box plots is to make a call about whether we are likely to see a difference in the **population**.

Case 1: If there is **no overlap** of the boxes, you **can** make the call that A **tends** to be larger than B in the **population**.

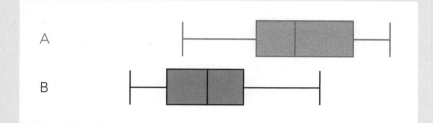

Case 2: If the median for one of the samples is outside the box for the other sample, then you **can** make the call that D **tends** to be larger than C in the **population**.

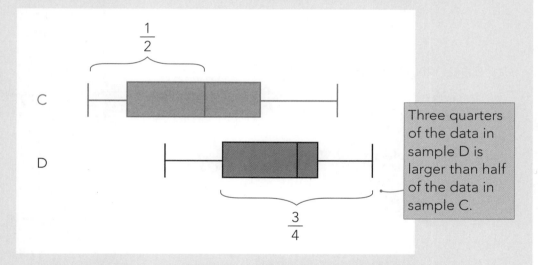

Three quarters of the data in sample D is larger than half of the data in sample C.

Case 3: If the above situations don't apply, then you must look at the difference between the medians and compare it to the overall visible spread.

OVS = **O**verall **V**isible **S**pread
DBM = **D**ifference **B**etween **M**edians

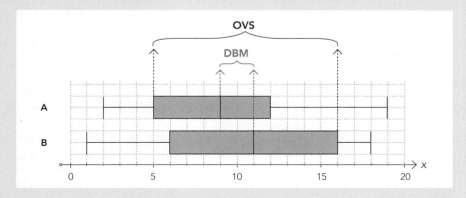

For a sample size of 30:
The **DBM** is more than **⅓ of the OVS** ⇒ individuals in population B tend to be bigger than those in population A.

For a sample size of 100:
The **DBM** is more than **⅕ of the OVS** ⇒ individuals in population B tend to be bigger than those in population A.

Method for sample size of 30:

Step 1: Draw lines to show the **OVS**:

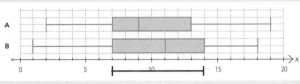

Step 2: By eye, divide the **OVS** into **thirds**:

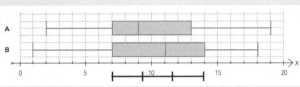

Step 3: Draw lines for the **DBM**:

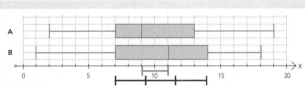

Step 4: Compare the **DBM** with the $\frac{OVS}{3}$.

*Because the **DBM** is less than the $\frac{OVS}{3}$, we **cannot** conclude that individuals in population B will tend to be longer than those in population A.*

Method for sample size of 100:

Step 1: Draw lines to show the **OVS**:

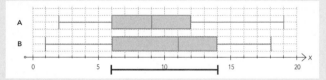

Step 2: By eye, divide the **OVS** into **fifths**:

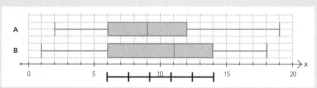

Step 3: Draw lines for the **DBM**:

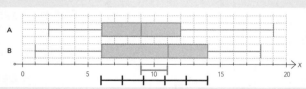

Step 4: Compare the **DBM** with the $\frac{OVS}{5}$.

*Because the **DBM** is more than the $\frac{OVS}{5}$, we **can** conclude that individuals in population B **will** tend to be longer than those in population A.*

In general: Do **not** measure or calculate these intervals. If you cannot tell that one is bigger by eye, then do **not** conclude that there is a difference in the population.

 ISBN: 9780170477437

Make the call and provide evidence for these box plots.

1 For a sample size of around 30:

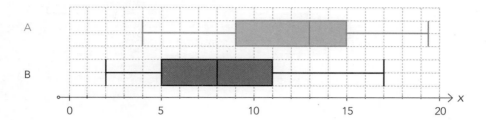

I notice: _____

This means:

☐ We **can** conclude that the individuals in population A tend to be bigger than those in population B.

☐ We **cannot** conclude that the individuals in population A tend to be bigger than those in population B.

2 For a sample size of around 30:

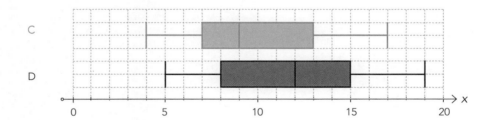

I notice: The DBM is/is not bigger than ⅓ of the OVS.

This means: We can/cannot conclude that the individuals in population A tend to be bigger than those in population B.

3 For a sample size of around 30:

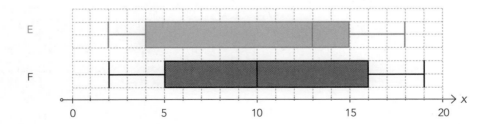

I notice: The DBM _____ bigger than ⅓ of the OVS.

This means: We _____ conclude that the individuals in population A tend to be bigger than those in population B.

4   For a sample size of around 100:

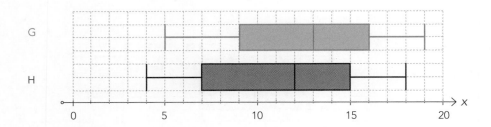

I notice: _____

This means: _____

_____

5   For a sample size of around 100:

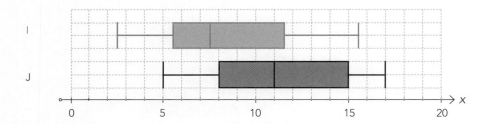

I notice: _____

This means: _____

_____

6   For a sample size of around 100:

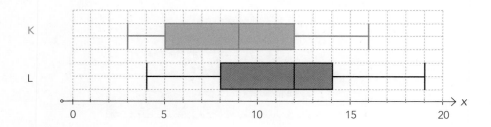

I notice: _____

This means: _____

_____

7   A sample of 94 students at Paradise High compared their travel time to get to school.

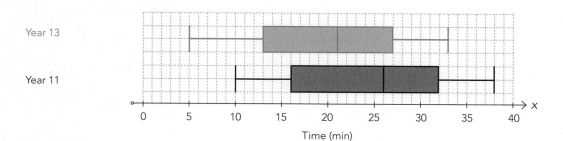

I notice: _____

This means: _____

_____

8   Latisha compared the lengths of dog walks her neighbours went on. Her sample size was 42.

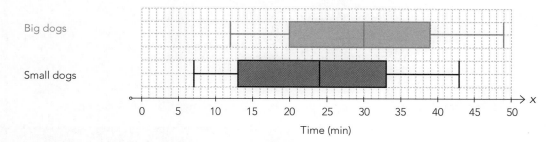

I notice: _____

This means: _____

_____

9   Gia recorded the lengths of 122 advertisements played by two radio stations.

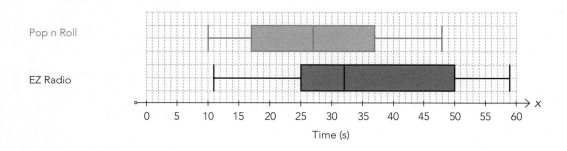

I notice: _____

This means: _____

_____

Dot plots with box plots

- Some statistical packages put a dot plot and a box plot on top of each other.
- Not all statistical features can been seen in both.

Jewelled gecko

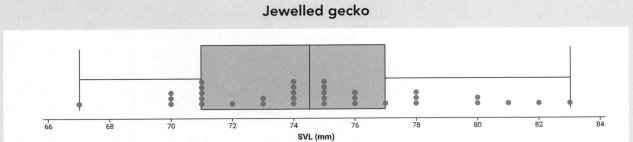

Tick which information can be seen on a dot plot and what can be seen on a box plot.

		Dot plot	Box plot
		✓ or ✗	✓ or ✗
Centre	Mode		
	Median		
	Mean		
Spread	Range		
	Interquartile range		
Shape			
Symmetry			
Unusual points and groupings			
The value of every piece of data			
Information for making the call			

Putting it all together

You will need to do the following:

1 Introduction
- Briefly **describe** the context of your investigation.
- Write or identify the investigative **question** or statement.
- Describe a **purpose** for your investigation.
- Identify the **populations**.

2 Investigation details

Collecting your own data	Given data
a Identify the **variables** you are collecting. **b** Consider and describe where your sample data is going to come from. **c** Describe in detail how you are **selecting** or **measuring** your data. **d** State how many pieces of data you collect.	**a** Identify the **variables** you have been given. **b** Describe where your sample data has come from. **c** State how many pieces of data you have.

3 Variation

Collecting your own data	Given data
Describe at least two different types of **variation**, and how you will manage these.	Describe at least two different types of **variation** that may have occurred, and your **assumptions** about how they have been managed.

4 Display your results
- Create a **dot plot** and a **box plot** to show your results.

5 Describe what you see in your graphs.
- Describe and compare the **centre**.
- Describe and compare the **spread**.
- Describe and compare the **shape**.
- Describe any **other features** such as unusual points or groupings.

6 Write a conclusion
- **Make the call. (Answer the question.)**
  — Are you able to conclude that there is a difference in the population?
- **Reflections**
  — How could you have improved the reliability of your results?
  — Are there other factors that you haven't mentioned so far that might have affected your results?
  — What are the next steps? What could this investigation lead to?

Annotated example

There are 48 species of gecko in New Zealand. The populations of both the orange-spotted and jewelled geckos are declining.
https://www.doc.govt.nz/nature/native-animals/reptiles-and-frogs/lizards/geckos/orange-spotted-gecko/

1   Introduction

Research is constantly occurring on these native geckos. A common measurement is the snout–vent length (SVL). This measurement (in millimetres) is often used when studying reptiles and can be used to compare species.

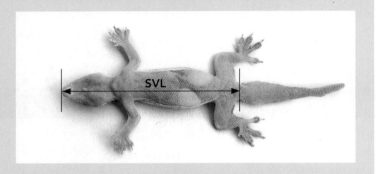

I wonder if the snout–vent length (SVL) of jewelled geckos tends to be longer than that of orange-spotted geckos in the South Island.

Question
Population
identified
Purpose

Knowledge of species' SVL measurements may give researchers insight into the habitat requirements/preferences of different species.

The two variables we are investigating are the snout–vent length (SVL) of orange-spotted and jewelled gecko. The data below is a random sample of 34 jewelled geckos and 46 orange-spotted geckos from the South Island.

2   Investigation
details

Given that the data was collected for us, we must assume that variation was managed appropriately by the researchers. Here are some potential sources of variation:

3   Variation

- Measurement variation — because the gecko were unlikely to have stayed still. We assume the researchers were trained in appropriate gecko-handling techniques to reduce stress and increase accuracy of measurement.
- Sampling variation — trapping success may be affected by the size of the gecko. Smaller geckos might be quicker and harder to catch. This may result in a sample that is not representative of the population.

4   Display

Dot plots and
box plots

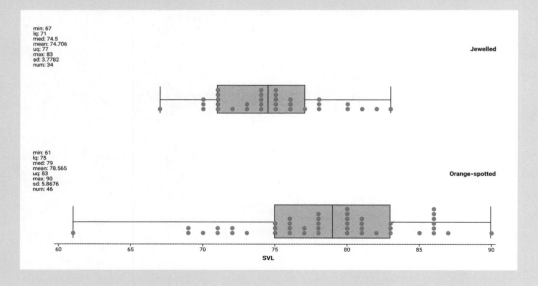

ISBN: 9780170477437

Centre

In the sample I notice that the median SVL of the orange-spotted geckos is 79 mm and the median SVL for the jewelled geckos is 74.5 mm. This is a difference of 4.5 mm. This means that in this sample, the median SVL for orange-spotted geckos tends to be longer than that for jewelled geckos.

Spread

The orange-spotted gecko SVL measurements (IQR = 8 mm) are more spread out than those of the jewelled gecko SVL (IQR = 6 mm). This indicates that the SVL data for orange-spotted geckos is more varied than that of the jewelled gecko.

Shape

In this sample, the distribution for the jewelled geckos' SVL tends towards a bell shape. The distribution for the orange-spotted geckos' SVL also tends towards a bell shape although it is slightly skewed to the left.

Unusual feature

There is one gecko that is away from the rest of the data with an SVL measurement of 61 cm.

This gecko's SVL is 8 cm smaller than the next smallest orange-spotted gecko. This piece of data is unusual and may have come from a juvenile orange-spotted gecko.

Conclusion — making the call

I notice that in this sample, 50% of the orange-spotted geckos' SVLs are longer than 75% of the jewelled geckos' SVLs.

This means that we can make the call that SVLs of orange-spotted geckos in the South Island will tend to be longer than the SVLs of the jewelled geckos.

If I took another sample, it's likely that I would get different results. However, I am confident that I would come to the same conclusion.

Reflections

* The conclusion would be more reliable if I had bigger samples.

* Geckos are known to be shy and difficult to catch. It might be that smaller geckos are harder to catch, and therefore not as well represented in the sample.

* I could investigate whether there is a difference between male and female geckos. It would also be interesting to compare these data with those of other species or with geckos from the North Island.

* I have assumed that the sample is representative of the population, that the geckos have been measured consistently, and the gecko species have been correctly identified and recorded.

Practice task

Introduction
The blue duck, or whio, is a species that is found only in New Zealand. Their numbers are estimated to be fewer than 3000. The data below is a sample of whio from the Tasman area.

Investigation details
Scientists collect anatomical data on this species from the Tasman region. The two variables we are investigating are the mass in grams and sex of whio from the Tasman area. We have a sample of 138 whio.

Question and purpose

_____

_____

_____

Variation

_____

_____

_____

_____

Whio

F

min: 560
lq: 750
med: 820
mean: 826.79
uq: 906.5
max: 1135
sd: 103.26
num: 161

Sex

min: 595
lq: 900
med: 985
mean: 975.63
uq: 1050
max: 1250
sd: 116.25
num: 151

M

Mass (g)

Centre

_____

_____

_____

_____

 ISBN: 9780170477437

Shape

_____

_____

_____

_____

Spread

_____

_____

_____

_____

Conclusion — make the call

_____

_____

_____

_____

_____

_____

_____

_____

_____

Reflections

_____

_____

_____

_____

_____

_____

_____

_____

_____

_____

Time series

- A time series is a series of values collected at **regular intervals** of time.
- The intervals could be years, quarters, months, weeks, days, hours, minutes, seconds or even milliseconds.
- These are plotted on a **line** graph using a graphing program, with time on the *x*-axis.
- The variable investigated must be **numeric** and may be discrete or continuous.

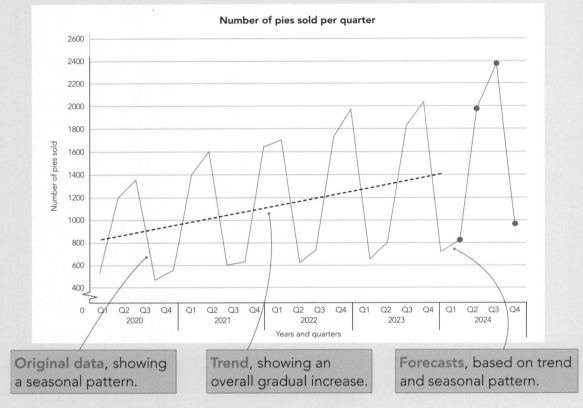

Original data, showing a seasonal pattern.

Trend, showing an overall gradual increase.

Forecasts, based on trend and seasonal pattern.

What's the point?
- The point is to analyse the **trend** and the **seasonal pattern** and use your analysis to **predict** what will happen in the future.
- The purpose of your statistical investigation should be clear and defined.
- Posing a question can help to clarify this.

What do we need to think about?
What is the population? What population is your inference about?

Type of data — you must have **one numerical** data set that is collected **over time**.

This can be discrete or rounded continuous data.

Sample size — you should have at least **five** complete cycles.

Cycles could be years, months or weeks.

Where am I getting my data?
If you are given it, where did it come from?

If you are collecting it, you must describe how you did this.

Variation
Explain at least two different types of variation and how these could be managed.

Understanding time series

What do the data and graphs look like?
Here are some examples (usually data sets will be larger).

Number of babies born in New Zealand named Florence

This data is yearly (annual).

Frequency means number of babies.

Time	Babies
2013	47
2014	43
2015	65
2016	61
2017	75
2018	78
2019	72
2020	89
2021	98
2022	87

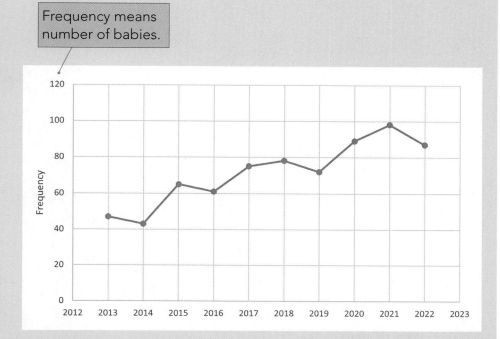

Number of marriages in New Zealand each quarter

This data is quarterly.

Time	Marriages
2020Q3	2577
2020Q4	5283
2021Q1	6189
2021Q2	3747
2021Q3	1701
2021Q4	4020
2022Q1	6759
2022Q2	3633
2022Q3	2601
2022Q4	5865
2023Q1	4782

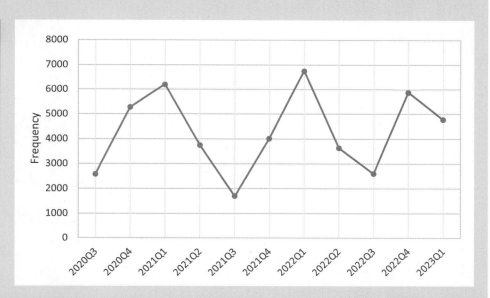

Quarter one is January, February and March.
Quarter two is April, May and June.
Quarter three is July, August and September.
Quarter four is October, November and December.

Number of babies born in New Zealand each month

This data is monthly.

Time	Babies born
2020M12	4803
2021M01	4743
2021M02	4797
2021M03	5373
2021M04	5169
2021M05	5385
2021M06	5079
2021M07	5166
2021M08	5142
2021M09	4995
2021M10	5268
2021M11	5178
2021M12	5010
2022M01	4842
2022M02	4608
2022M03	5016
2022M04	4635
2022M05	4875
2022M06	4449
2022M07	4665
2022M08	4719
2022M09	4602
2022M10	4803
2022M11	4482
2022M12	4594

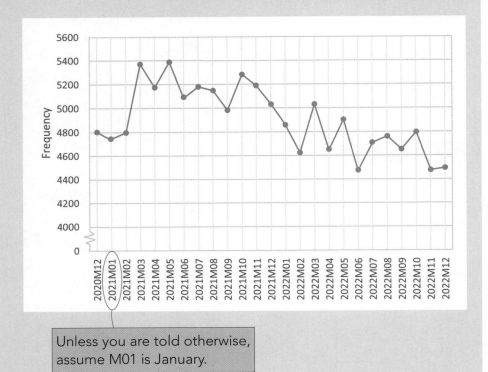

Unless you are told otherwise, assume M01 is January.

Power usage in a New Zealand house each day

This data is daily.

Time	Power
Week1D7	37.51
Week2D1	26.71
Week2D2	27.73
Week2D3	36.3
Week2D4	28.78
Week2D5	29.23
Week2D6	54.79
Week2D7	65.41
Week3D1	25.1
Week3D2	33.13

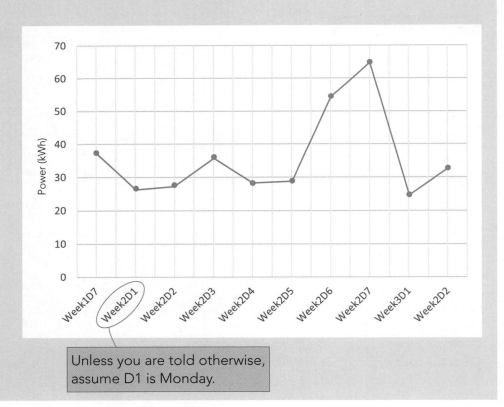

Unless you are told otherwise, assume D1 is Monday.

Features of time series

1 The trend

- Trend lines can be added in most computer programs.
- These display any overall change. This could be **increasing**, **decreasing**, **stable** or **inconsistent**.
- Some programs will quantify the trend values for you.

Increasing trend

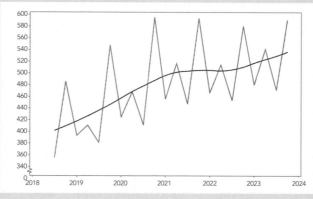

Decreasing trend

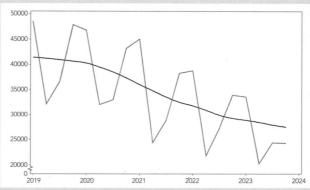

Stable trend

There are no significant increases or decreases.

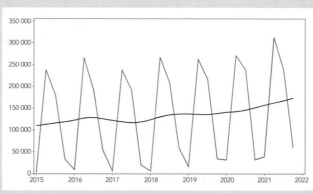

Inconsistent trend

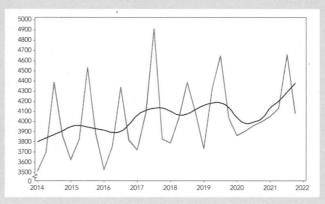

Highlight the description that best fits the overall trend of these graphs.

1

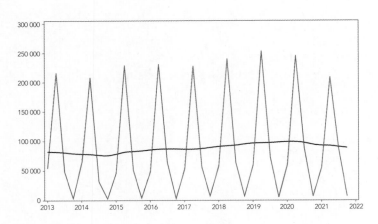

Increasing

Decreasing

Stable

Inconsistent

2

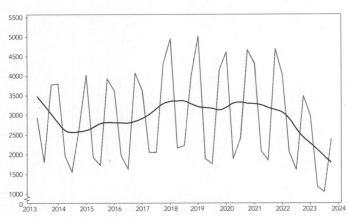

Increasing

Decreasing

Stable

Inconsistent

3

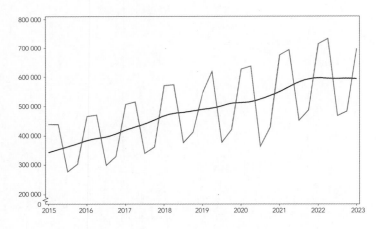

Increasing

Decreasing

Stable

Inconsistent

4

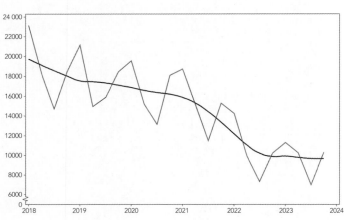

Increasing

Decreasing

Stable

Inconsistent

ISBN: 9780170477437

Inconsistent trends, important bits

Trends are often inconsistent, so it's important to focus just on the significant changes rather than the detail.

Example:

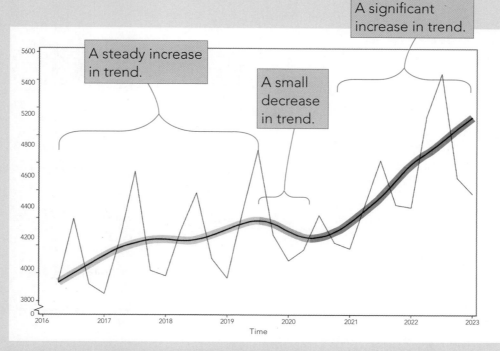

There is an overall increase in the trend from early 2016 to the end of 2022.
Between 2016 and mid-2019, there was a steady increase in the trend.
From mid-2019 until mid-2020, there was a slight decrease in the trend.
From mid-2020 onwards, there was a significant increase in the trend.

Answer the following questions.

5

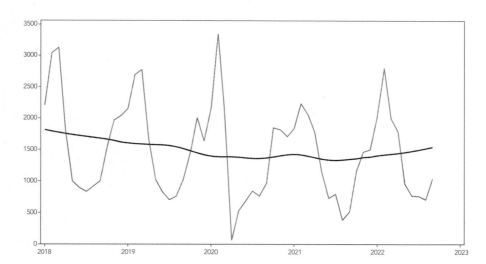

a Overall, the trend is increasing/decreasing/stable/inconsistent from the start of

_____ to the end of _____.

b From 2018 until mid-2021, the trend is _____.

c From mid-2021 until mid-2022, the trend is _____.

6

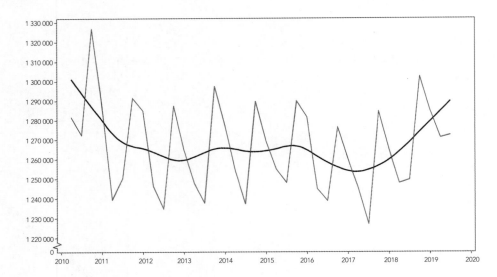

a Overall, the trend has _____ from the start of _____ to

the end of _____.

b If you said that overall the trend had decreased, does this tell the whole story?   Yes/No

Why/why not? _____.

c Between 2010 and 2013, the trend was _____.

d The trend was stable from _____ until _____.

e From 2017 until mid-2019, the trend was _____.

7

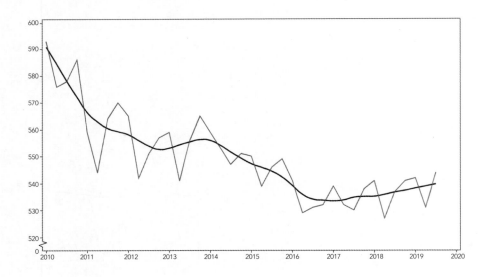

a Overall, the trend has _____ from the start of _____ to

the end of _____.

b The largest decrease in the trend was between _____ and _____.

c What evidence is there to support this? _____

_____

Quantifying and calculating the change in the trend

- It is important to quantify the amount that the trend has changed over time.
- First quantify the overall change, then calculate an average change per month, quarter, year, etc.

Make an overall statement:

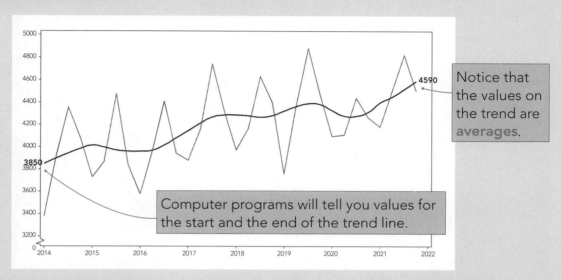

Notice that the values on the trend are averages.

Computer programs will tell you values for the start and the end of the trend line.

I notice that overall, there has been an increasing trend from an average of 3850 units per quarter at the start of 2014 to 4590 units per quarter at the end of 2021.
This means that over this period, there has been an increase of 740 units.

This represents an average increase of $\dfrac{740}{8}$ = 92.5 units per year.

Over the 8 years, the trend has increased by 740 units.

Answer the following questions.

8

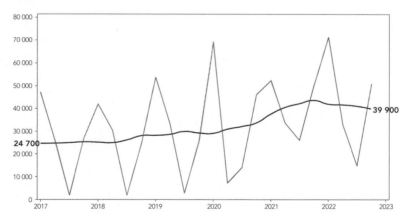

a **I notice** that overall, the trend has increased/decreased from an average of _____

units per quarter at the start of 2017 to _____ units per quarter at

the end of 2022.

b **This means** that over this period, the increase/decrease is

_____ − _____ = _____.

c **This represents** an average increase/decrease of _____ per year.

9

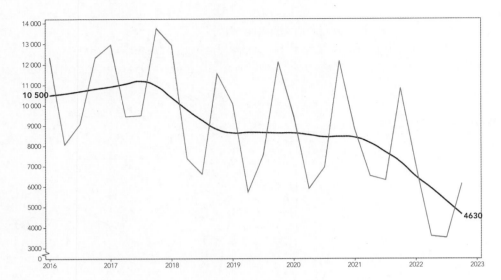

a   **I notice** that overall, the trend has increased/decreased from an average of

_____ units per quarter at the start of 2016 to _____ units per quarter at

the end of 2022.

b   **This means** that over this period, the increase/decrease is

_____ – _____ = _____.

c   **This represents** an average increase/decrease of _____ per year.

10

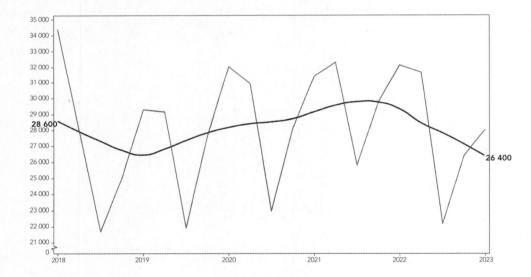

a   **I notice** _____

_____.

b   **This means** _____

_____.

c   **This represents** _____.

ISBN: 9780170477437

Putting it together with context

This graph shows the price per kg of capsicums for each month in New Zealand from the start of 2018 until the end of 2022.

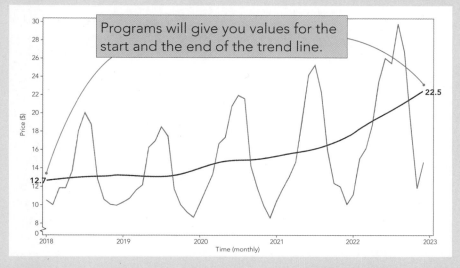

Make an overall statement:

I notice that overall, there has been an increasing trend in the average price per kg of capsicums from **$12.70** at the start of 2018 to **$22.50** at the end of 2023. There was an initial plateau, then a steady increase after 2020.

This means that over this period, the price per kg of capsicums has increased by $9.80.

This represents an average increase in the price per kg of capsicums of $1.96 per year.

Answer the following questions.

11 This graph shows the number of earthquakes of strength 4 or more on the Richter scale each year in New Zealand.

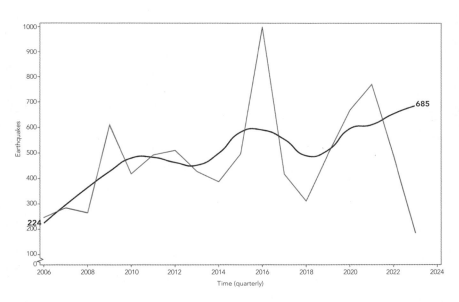

a **I notice** that overall, the trend for the average number of earthquakes of strength 4 or more

per quarter in New Zealand each year has been increasing/decreasing, from _____

at the start of 2006 to _____ at the end of 2023.

b **This means** that over this period, the number of earthquakes has increased/decreased by

_____ .

c **This represents** an average increase/decrease of _____ earthquakes per year.

12 This graph shows the number of pairs of twins born in New Zealand each year.

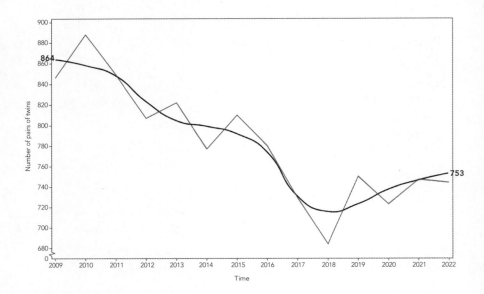

a I notice _____

_____.

b This means _____

_____.

c This represents _____.

13 This graph shows the price per kg of avocados in New Zealand.

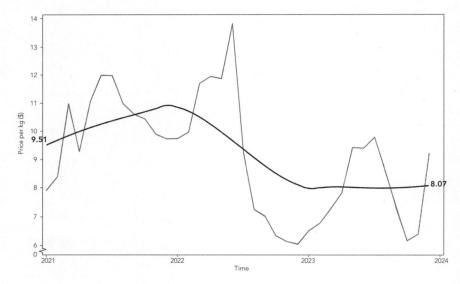

a I notice _____

_____.

b This means _____

_____.

c This represents _____.

2 Repeating patterns

- Some time series have a repeating pattern, while others do not.
- These repeating patterns usually occur at regular intervals along the time series.
- These may be due to: annual changes, e.g. seasonal change in the weather
  weekly changes, e.g. effects of weekends on traffic flows
  daily changes, e.g. effects of light on photosynthesis in plants.

A time series without a pattern

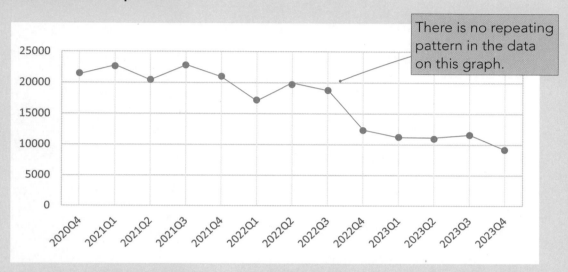

There is no repeating pattern in the data on this graph.

A time series with a repeating pattern

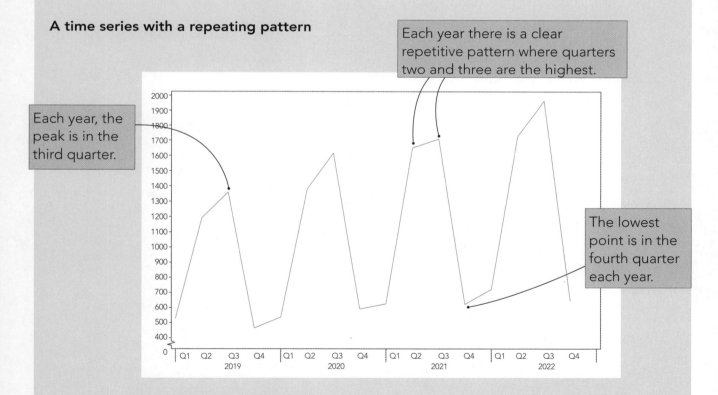

Each year, the peak is in the third quarter.

Each year there is a clear repetitive pattern where quarters two and three are the highest.

The lowest point is in the fourth quarter each year.

Notice that the pattern isn't identical each year.

In a report, you should identify the time periods that have the highest (peaks) and lowest (troughs) points.

You should also try to find the reasons for these.

Complete the following questions.

1

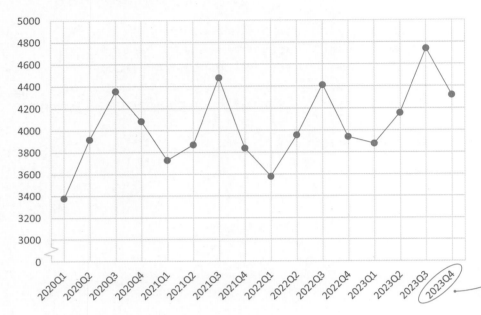

2023Q4 means the fourth quarter of 2023.

a The time interval between each data point is days/weeks/months/quarters/years.

b The pattern is monthly/quarterly/yearly.

c Highlight the four peaks. They occur in quarter ＿＿＿＿＿＿＿.

d The lowest points of the pattern is in quarter ＿＿＿＿＿＿＿.

2

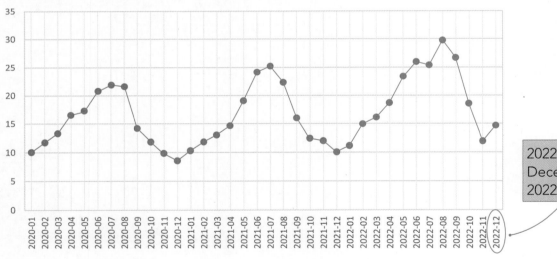

2022-12 means December of 2022.

a The time interval between each data point is days/weeks/months/quarters/years.

b The pattern is monthly/quarterly/yearly.

c Highlight the four peaks. They occur in ＿＿＿＿＿＿＿＿＿＿＿＿＿＿＿＿＿＿＿＿＿.

d The lowest points of the pattern is in ＿＿＿＿＿＿＿＿＿＿＿＿＿＿＿＿＿＿＿＿＿.

Patterns in context

- You can add detail to your seasonality statements with context.

Example: This graph shows the prices of apples per kg each month in New Zealand.

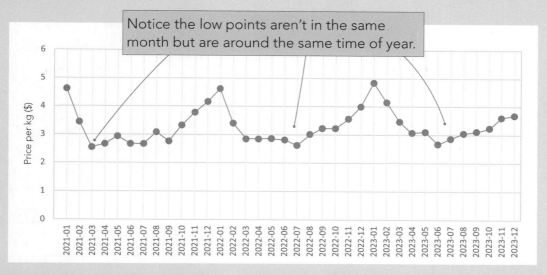

Notice the low points aren't in the same month but are around the same time of year.

I notice there is a recurring ~~weekly/monthly/quarterly/yearly~~ pattern with a peak of about **$4.80** in **January** and the minimum of **$2.50 to $3.00** in any month **between March and July**.

This means the apples are about $2 per kg more expensive in January than in the winter months.

Reason: Apples are ripe in autumn, so they are most abundant (and cheapest) during the autumn and winter months. By January, there aren't so many and therefore they are more expensive.

Answer the following questions.

3 This graph shows the number of sunshine hours each quarter in Tauranga.

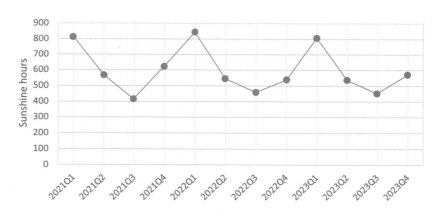

a I notice there is a recurring weekly/monthly/quarterly/yearly pattern with a peak of about

_____ in _____ and a minimum of about _____ in

_____.

b This means the highest number of sunshine hours in Tauranga usually occur in

_____ and the lowest in _____.

c Reason: _____

_____

4 This graph shows the price of cucumbers each month in New Zealand.

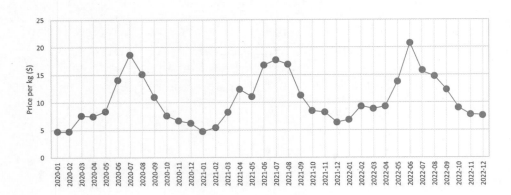

a I notice _____

_____

_____.

b This means _____

_____.

c Reason: _____

_____

5 This graph shows the number of visitors to an arcade over three weeks.

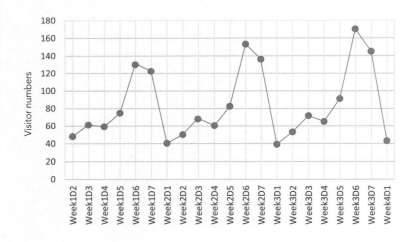

_____

_____

_____

_____

_____

3 Variation in the data

- Variation refers to how far away the data is from the trend line.
- This may increase, decrease, be consistent or be irregular.

Increasing variation

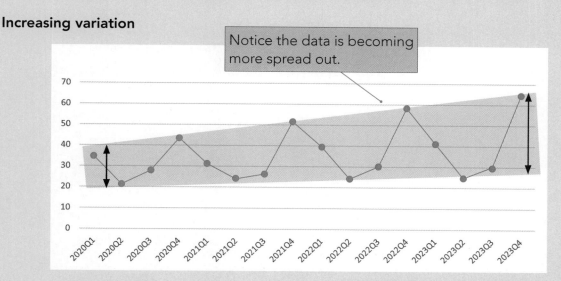

Notice the data is becoming more spread out.

Decreasing variation

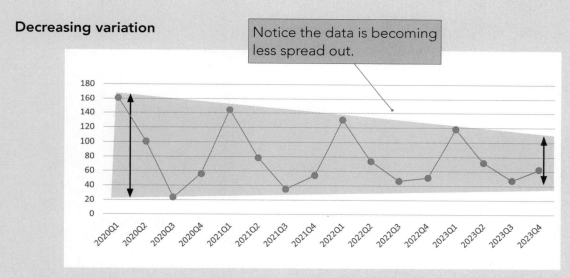

Notice the data is becoming less spread out.

Constant variation

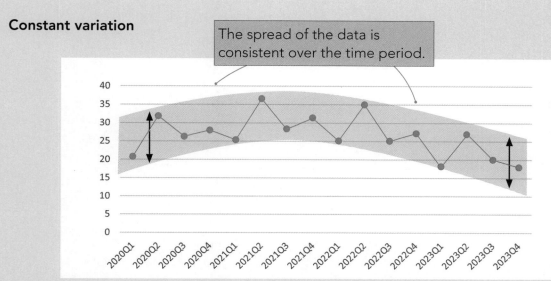

The spread of the data is consistent over the time period.

Highlight the best description of the variation in these data sets.

1

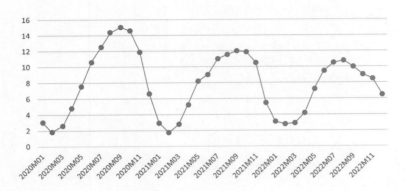

Increasing

Decreasing

Constant

Irregular

2

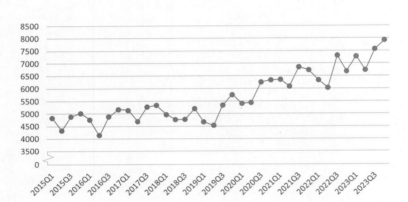

Increasing

Decreasing

Constant

Irregular

3

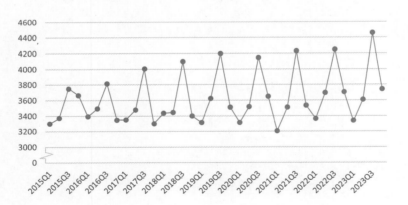

Increasing

Decreasing

Constant

Irregular

4

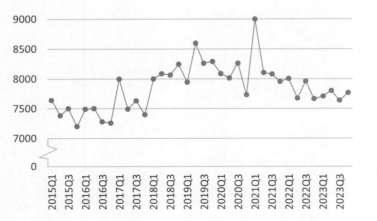

Increasing

Decreasing

Constant

Irregular

ISBN: 9780170477437

5

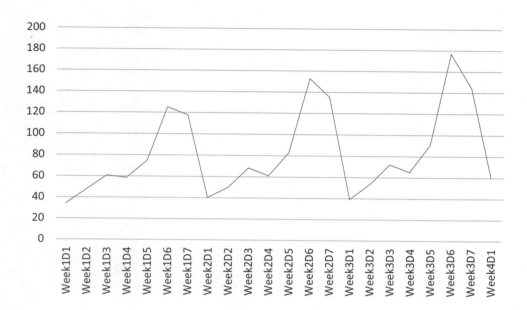

a   I notice that the variation in the data is increasing/decreasing/constant/irregular.

b   This means that the data is _____.

6   This graph shows the amount of wine (millions of litres) exported between the start of 2015 and the third quarter of 2023.

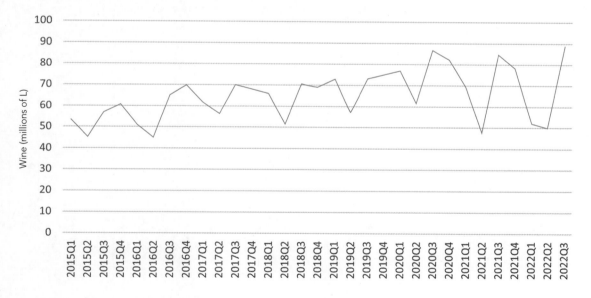

a   I notice that the variation in the data between the start of 2015 until middle of 2021 is increasing/decreasing/constant/irregular, but after that it increases/decreases, and becomes constant/irregular.

b   This means _____

_____

_____.

4 Unusual points

- These are points which lie some distance from where we would expect.
- These may be valid points or could be the result of measurement error or recording error, so they should be checked if possible.
- It is important to identify unusual points. You may or may not be able to find a reason for them.

Example:

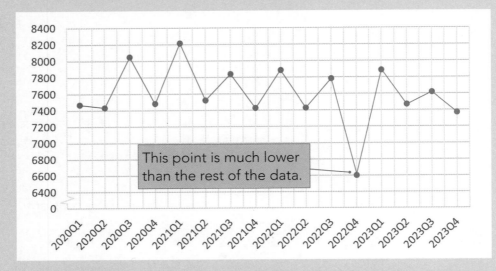

This point is much lower than the rest of the data.

In the fourth quarter of 2022, there is an unusually low value of 6600 units. I would have expected this to be closer to 7400 units.

You may need to look at the original data to get more detail about the point.

Identify the unusual point(s) and complete the sentence.

1

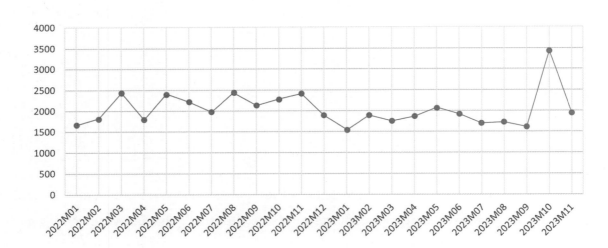

I notice that _____

_____.

 ISBN: 9780170477437

2

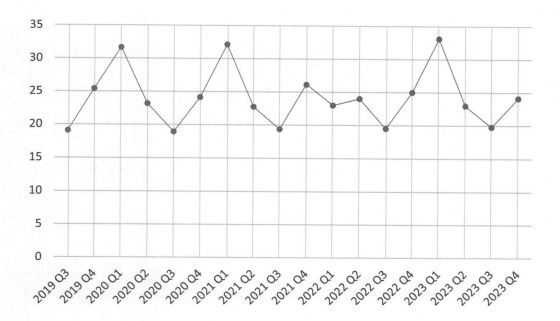

I notice that _____

_____.

3

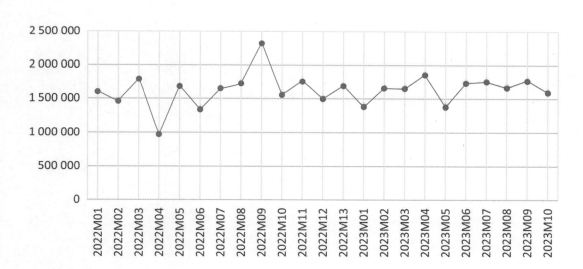

I notice that _____

_____.

Forecast or prediction

- We can use historical data and the trend line and the pattern to make forecasts or predictions.
- Because we are using mathematical models to describe real situations, we can never be certain that predictions will be accurate.
- How accurate they are will depend on:
  — how consistent the data has been
  — how much data has been collected.
- Any prediction into the future will be uncertain, and the further into the future you predict, the more uncertainty there will be around the prediction.

Example:

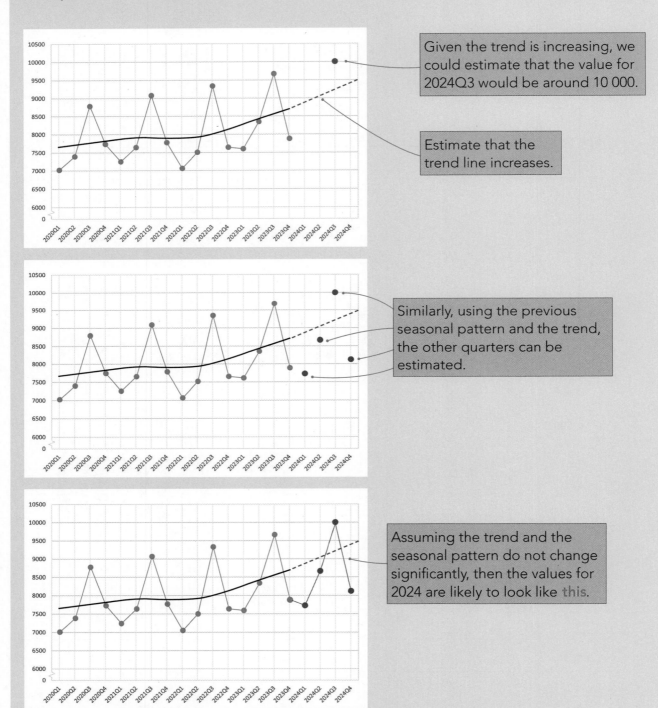

Given the trend is increasing, we could estimate that the value for 2024Q3 would be around 10 000.

Estimate that the trend line increases.

Similarly, using the previous seasonal pattern and the trend, the other quarters can be estimated.

Assuming the trend and the seasonal pattern do not change significantly, then the values for 2024 are likely to look like this.

 ISBN: 9780170477437

Answer the following questions.

1

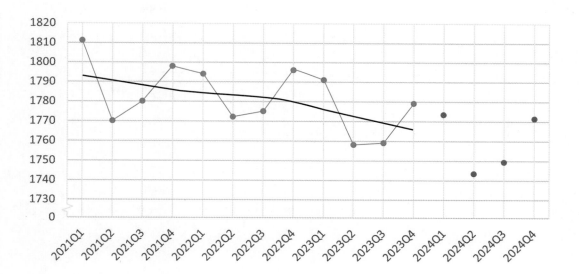

a Continue the trend line.

b Estimate the values for:

2024Q1 _____ 2024Q2 _____ 2024Q3 _____ 2024Q4 _____

c Rule lines between the points.

2

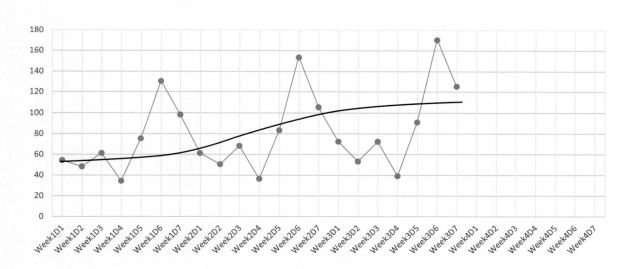

a Continue the trend line.

b Estimate the values for:

Week4D1 _____ Week4D2 _____ Week4D3 _____ Week4D4 _____

Week4D5 _____ Week4D6 _____ Week4D7 _____

c Rule lines between the points.

3 This graph shows the average quarterly temperatures for Dunedin.

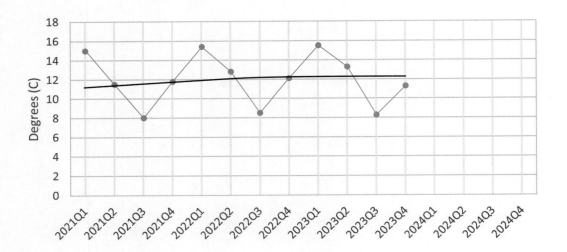

a Continue the trend line.

b Estimate the temperatures for Dunedin for:

2024Q1 _____ 2024Q2 _____ 2024Q3 _____ 2024Q4 _____

c Rule lines between the points.

4 This graph shows the number of sunshine hours per quarter in Wellington.

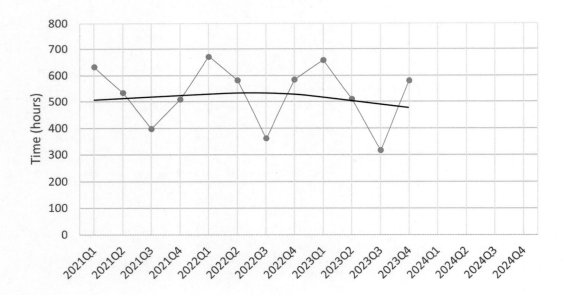

a Continue the trend line.

b Estimate the sunshine hours for Wellington for:

2024Q1 _____ 2024Q2 _____ 2024Q3 _____ 2024Q4 _____

c Rule lines between the points.

 ISBN: 9780170477437

Confidence in your predictions

- Your confidence in the accuracy of your predictions will depend on:
  — the consistency of the trend
  — the consistency of the pattern
  — the amount of variation
  — how far into the future the prediction is for and
  — the amount of data that has been collected (at least five cycles).

Indicate with ticks or crosses whether each feature of these time series is likely to lead to confident predictions. Overall, would you be likely to be able to make accurate predictions?

5

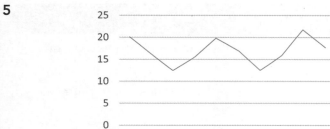

Consistent trend	
Consistent pattern	
Consistent variation	
Enough data	
Accurate predictions?	Yes/No

6

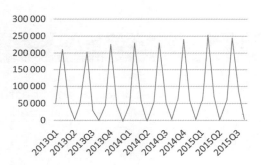

Consistent trend	
Consistent pattern	
Consistent variation	
Enough data	
Accurate predictions?	Yes/No

7

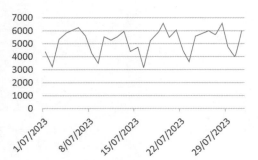

Consistent trend	
Consistent pattern	
Consistent variation	
Enough data	
Accurate predictions?	Yes/No

8

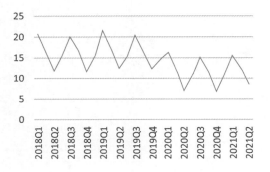

Consistent trend	
Consistent pattern	
Consistent variation	
Enough data	
Accurate predictions?	Yes/No

Putting it all together

You will need to do the following:

1 Introduction

- Briefly **describe** the context of your investigation.
- Write or identify the investigative **question** or statement.
- Describe a **purpose** for your investigation.

2 Investigation details

Collecting your own data	Given data
a Identify the **variables** you are collecting. **b** Consider and describe where your sample data is going to come from. **c** Describe in detail how you are **selecting** or **measuring** your data. **d** State how many pieces of data you collect.	**a** Identify the **variables** you have been given. **b** Describe where your sample data has come from. **c** State how many pieces of data you have.

3 Variation

Collecting your own data	Given data
Describe at least two different types of **variation**, and how you will manage these.	Describe at least two different types of **variation** that may have occurred, and your **assumptions** about how they have been managed.

4 Display your results

- Create a **time series graph**, including a **trend line**.

5 Describe what you see in your graphs.

- Describe what you see in the **trend**.
- Describe any **patterns**.
- Describe any **other features**.
- Estimate a **prediction** for a future time period.

6 Write a conclusion

- Do you think your prediction is reliable?
- **Reflections**
  — How could you have improved the reliability of your results?
  — Are there other factors that you haven't mentioned so far that might have affected your results?
  — What are the next steps? What could this investigation lead to?

ISBN: 9780170477437

Annotated example

Auckland Zoo has the largest diversity of wildlife in New Zealand. It is home to over 2800 animals, with at least 130 different species. It has had over 28 million visitors since opening in 1922. https://www.aucklandzoo.co.nz/

The data below shows the number of daily visitors for the month of September 2022. I wonder if we could use this data to predict future visitor numbers?
The zoo needs to know how many people are likely to visit each day so that they can plan their staffing schedules, food supplies for their café, etc.

Given that the data was collected for us, we must assume that variation was managed appropriately by the collectors. Here are some potential sources of variation:
- Measurement variation could have occurred with inaccurate visitor numbers recorded. If they used financial information to record these numbers, a two-for-one visitor pass could result in an error.
- Occasion-to-occasion variation — visitor numbers could be affected by public holidays or the weather.

Note: 1st of September was a Thursday.

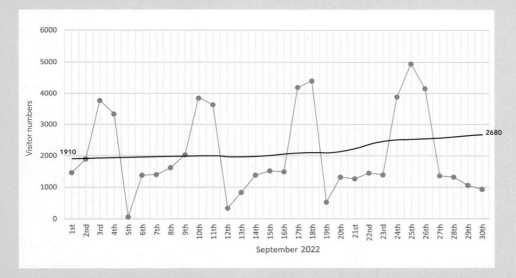

Trend
There is an overall increase in the trend from 1910 visitors each day at the start of September to 2680 visitors each day towards the end of September 2022. Over the 30 days, the number of visitors has increased by 770. The average increase in the number of visitors is 26 (25.7 to 1 dp) per day. When visiting a zoo, most time is spent outside, so this increase is probably due to improving weather during the month.

Pattern
I notice that there is a recurring weekly pattern in the data. There are clear peaks on Saturdays and Sundays, which all have over 3000 visitors. Apart from the 26th of September, these are all weekend days. The lowest points are on the 5th, 12th and 19th of September. These are all Mondays.
This means that the zoo has the highest number of visitors in the weekend, when it can expect between 3500 and 5000 people. It generally has the lowest number of visitors (fewer than 500) on Mondays.

Variation in the data

There is a fairly constant variation in the number of visitors throughout the week, with about a 4000 visitor difference between the busy and quiet days.

Unusual points

I notice one usual point on Monday the 26th of September. There are around 4000 visitors when we would normally expect around 400. This is because it was a public holiday for New Zealanders to celebrate and mourn Queen Elizabeth's life and death.

Prediction

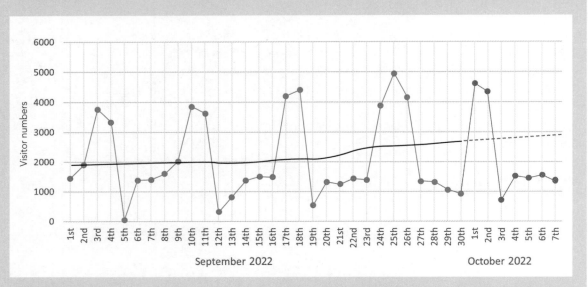

I estimate that the number of visitors to the Auckland Zoo on the first weekend of October will be around 4500 each day. On Monday the 3rd of October, the zoo can expect around 800 visitors. During the rest of the week, it is likely to get around 1500 visitors each day.

Conclusion

There is a consistent slightly increasing trend and a consistent weekly pattern. I think Auckland Zoo can be reasonably confident of these predictions. This information should help managers to staff the zoo and to plan adequately for the next few months.

Reflections

It would be interesting to find out how visitor numbers vary throughout the year, and also to investigate the effect of weather on numbers of visitors. Because visiting a zoo is mostly an outdoor activity, I would expect weather to significantly influence the number of visitors each day.

ISBN: 9780170477437

Practice task

Introduction

There are 28 cycle counters positioned around Christchurch. These counters record when a cyclist goes past. This gives a snapshot of cycle use in the city at various points.

Investigation details

Data from the cycle counter at North Hagley Park was recorded from 1st July to 4th August 2023.
https://ccc.govt.nz/transport/improving-our-transport-and-roads/traffic-count-data/cycle-counters

Question and purpose

_____

_____

_____

Variation

_____

_____

_____

_____

Note: 1st July was a Saturday.

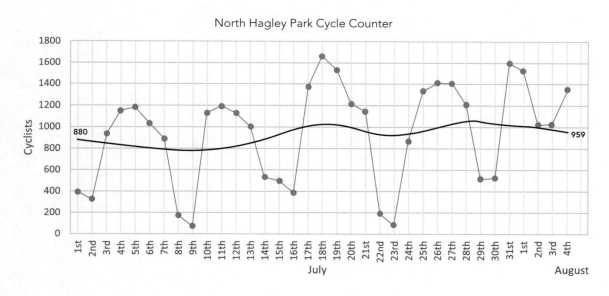

Trend

_____

_____

_____

_____

Cycle

_____

_____

_____

_____

Variation in the data

_____

_____

_____

_____

Unusual points

_____

_____

_____

_____

Predictions

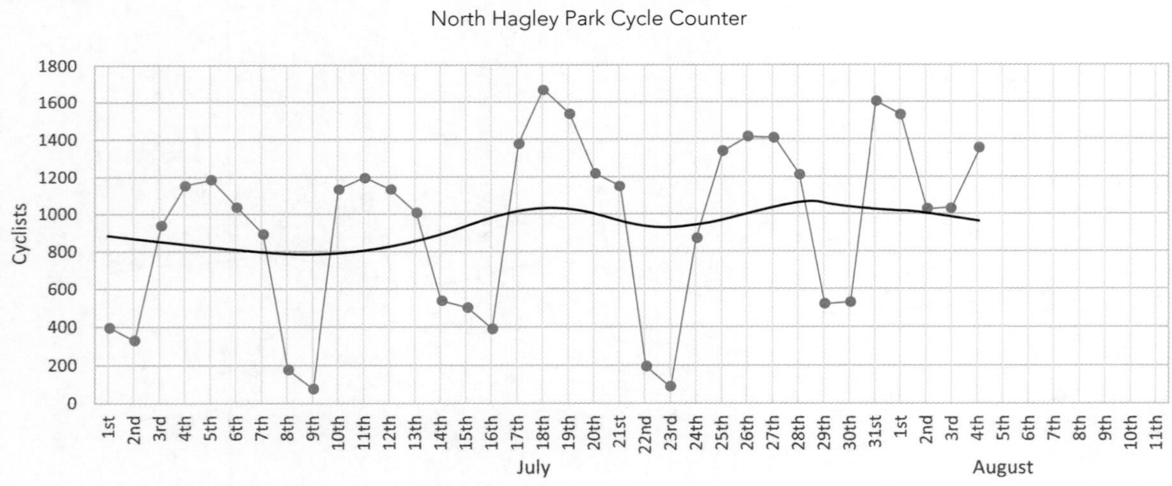

North Hagley Park Cycle Counter

_____

_____

_____

_____

_____

Conclusions

_____

_____

_____

_____

_____

_____

Reflections

_____

_____

_____

_____

_____

_____

Experimental probability

Probability

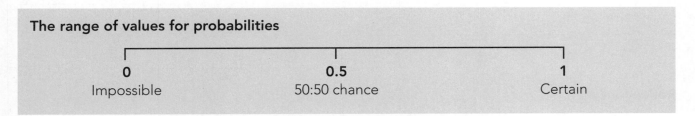

The range of values for probabilities

0	0.5	1
Impossible	50:50 chance	Certain

Discuss the meanings of the following words and phrases with others in your class, and match each to appropriate probability value(s). You may have several words beside the same probability value, and some words can cover several probability values.

> very unlikely, improbable, slight chance, a sure thing, extremely likely,
> certain, maybe, likely, fifty-fifty, impossible, possible, probable,
> unlikely, no chance, definite, even chance, no way, very likely, maybe

0	
0.1	
0.2	
0.3	
0.4	
0.5	
0.6	
0.7	
0.8	
0.9	
1.0	

 ISBN: 9780170477437

Using numbers for writing probabilities

- Probabilities can be written as **fractions**, **decimals** or **percentages**.
- You can convert between these with your calculator.
- When you want to **compare** probabilities, you should always use **decimals**.

Example: Write the probability $\frac{5}{8}$ as a decimal.   5 8 0.625

Convert the following probabilities to decimals, and state which of each pair is more likely.

1 $\frac{3}{8} =$ _____ $\frac{2}{5} =$ _____

More likely: _____

2 $\frac{2}{3} =$ _____ $\frac{7}{11} =$ _____

More likely: _____

3 $\frac{17}{20} =$ _____ $\frac{22}{25} =$ _____

More likely: _____

4 $\frac{5}{18} =$ _____ $\frac{1}{4} =$ _____

More likely: _____

5 $\frac{1}{8} =$ _____ $\frac{2}{15} =$ _____

More likely: _____

6 $\frac{7}{16} =$ _____ $\frac{4}{9} =$ _____

More likely: _____

Test yourself

For the following sentences, indicate whether each statement could be correct or is impossible. Justify your decision.

✔ or ✖

7 Suzie thinks her chance of passing her driving licence on her first attempt is about –0.2.

_____

8 The probability that Teina's football team wins their game is $\frac{2}{3}$.

_____

9 Noa is goal shoot in her netball team. In her current form, the probability that she scores on an attempt is 70%.

_____

10 Isaac says he has a 110% chance of being picked for the A rugby team.

_____

Ways of calculating probabilities

1 Equally likely outcomes

- This is where every possible outcome is **equally likely** to occur.
- An example is throwing a fair die: you are equally likely to get a 1, 2, 3, 4, 5 or 6.

$$\text{probability} = \frac{\text{number of favourable outcomes}}{\text{total possible outcomes}}$$

With a single event

Examples:

1 For a fair die:

$$P(6) = \frac{1}{6} \text{ or } 0.1\dot{6} \qquad\qquad P(1 \text{ or } 2) = \frac{2}{6} \text{ or } \frac{1}{3} \text{ or } 0.\dot{3}$$

2 For the spinner shown in the diagram:

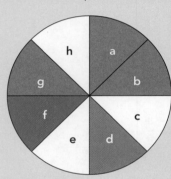

$$P(d) = \frac{1}{8} \text{ or } 0.125$$

$$P(grey) = P(d \text{ or } f) = \frac{2}{8} \text{ or } 0.25$$

> Notice that **a** is both green and a vowel, but we include it just **once** in the list.

$$P(\text{green or vowel}) = P(a, b, e \text{ or } g) = \frac{4}{8} \text{ or } 0.5$$

$$P(\text{white and vowel}) = P(e) = \frac{1}{8} \text{ or } 0.125$$

$$P(\text{grey and } h) = \frac{0}{8} \text{ or } 0$$

1 Calculate the following probabilities.
When tossing a fair die:

a P(2) = _____

b P(1, 5 or 6) = _____

c P(odd number) = P(1, 3, 5) = _____

d P(not a 3) = P(_____) = _____

e P(6 or an odd number) = P(_____) = _____

f P(7) = _____

g P(1, 2, 3, 4, 5 or 6) = _____

2 Anna created a different spinner:

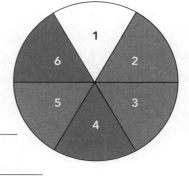

a P(4) = _____

b P(4 or 6) = _____

c P(green) = P(_____) = _____

d P(not green) = P(_____) = _____

e P(7) = _____

f P(odd number or green) = P(_____) = _____

g P(odd number and green) = P(_____) = _____

3 Mike has a bag of lollies which are all the same apart from their colour. It contains 5 red lollies, 8 green ones, 6 yellow ones and 1 white one. He puts his hand in the bag and selects a lolly at random. After he has looked at it, he puts it back in the bag.

a P(red) = _____

b P(not getting a red) = _____

c P(red or green) = _____

d P(not getting a red or green) = _____

e P(white, red or green) = _____

f P(not yellow) = _____

g P(black) = _____

h P(red, green, yellow or white) = _____

4 Fetu has a set of dominoes. Each domino has two halves, with between zero and six spots at each end. There are 28 dominoes in a full set. He places all his dominoes face down on the floor, and turns one over at random. After he has looked at it, he turns it over and mixes the dominoes up. Give your answers as fractions.

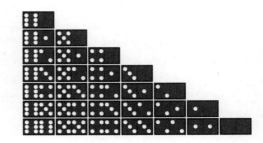

a P(a domino with 12 dots on it) = _____

b P(a domino with the same number of dots at each end) = _____

c P(a domino with six at one end and exactly three or four at the other end) = _____

d P(a domino which has at least four dots on one end and none on the other) = _____

e P(a domino with a *total* of six dots on it) = _____

2 Experimental probability

- Sometimes the probability of an event is **difficult or impossible** to calculate.
- In these cases, we do an experiment and perform **many trials** and record the number of times an event occurs.
- Often, the true value of the probability will **never** be known.
- **The greater the number of trials, the closer our estimate for the probability will be to the true probability.**

$$\text{probability} = \frac{\text{number of times an event occurs}}{\text{total number of trials}}$$

Example: Students were asked to keep a record of where their lunches came from during the first term. Erin had no lunch on 12 days, brought it from home on 21 days, and bought it from the canteen on 9 days and from the dairy on 6 days. Calculate the following probabilities.

$$\text{P(she had no lunch)} = \frac{\text{number of days with no lunch}}{\text{total number of days in the term}} = \frac{12}{48} = \frac{1}{4} \text{ or } 0.25$$

$$\text{P(she brought lunch from home)} = \frac{\text{number of days with lunch brought from home}}{\text{total number of days in the term}}$$
$$= \frac{21}{48} = \frac{7}{16} \text{ or } 0.4375$$

$$\text{P(she bought lunch at the canteen)} = \frac{\text{number of days with lunch from the canteen}}{\text{total number of days in the term}}$$
$$= \frac{9}{48} = \frac{3}{16} \text{ or } 0.1875$$

$$\text{P(she bought lunch at the dairy)} = \frac{\text{number of days with lunch from the dairy}}{\text{total number of days in the term}} = \frac{6}{48} = \frac{1}{8} \text{ or } 0.125$$

Answer the following questions. Round your answers to 2 dp.

1 Alfie was in the same class as Erin, and he also kept records of where his lunches came from.
- He had lunch every day.
- He brought it from home on 26 days.
- He bought it from the canteen on 13 days.
- He bought it from the dairy on the remaining days.
Calculate the probability for each of these options.

_____

_____

_____

2 There are 28 students in a class. Each person wrote down their favourite winter sport. For 10 it was rugby, 7 chose basketball, 4 chose hockey, 3 chose netball, 3 chose badminton and for 1 student it was curling. Calculate the probabilities that a student chose each sport.

_____

_____

_____

3 A survey of the time spent watching television each school day by 160 secondary school students produced the following results.

Time watching television	Number of students
None	32
$0 <$ time ≤ 1	51
$1 <$ time ≤ 2	20
$2 <$ time ≤ 3	22
Over 3 hours	35

a What is the probability that a student watches no television?

b What is the probability that a student watches one hour or less of television each school day?

c What percentage of students watch more than two hours of television each school day?

4 The same survey asked students about the number of pets in their homes.

Number of pets	Number of students
0	37
1	49
2	31
3	24
4 or more	19

a What is the probability that a student's home has a pet?

b What is the probability that a student's home has just two or three pets?

c What percentage of students have exactly one pet?

5 The same survey asked students how they had got to school that morning.

How they got to school	Number of students
Walked	58
Scooter or skateboard	26
On a bus	
Motor bike or motor scooter	7
By car — drove themselves	15
By car — driven by somebody else	23

a How many students took a bus to school?

b What is the probability that a student walked to school?

c What is the probability that a student came to school by car?

Expected number of outcomes

expected number of outcomes = P(event) x number of trials

Examples:

1 Amira throws a die 210 times. How many times can she expect a one?

Expected number of ones = P(1) x 210

$$= \frac{1}{6} \times 210$$

$$= 35$$

2 The probability of being struck by lightning in the New Zealand is $\frac{1}{280\,000}$ in one year. If in 2023 the population of the New Zealand is 5 199 000, calculate the expected number of people who will be struck by lightning during 2023.

Expected number of people struck by lightning = $\frac{1}{280\,000}$ x 5 199 000

$$= 18 \text{ or } 19$$

The exact answer is 18.56… but this is people, so we must round to the nearest whole person. If the decimal is around a half, then round up and down.

Answer the following questions.

1 Hemi flips a coin 50 times and records the results. How many 'heads' can he expect?

2 Nikau tossed a die 80 times and recorded the results. How many times can he expect to get a one or a two?

3 The probability of being left handed is 0.1. If your school roll is 654, how many people would you expect to be left handed at your school?

4 The probability that a boy is colour blind is 0.08. If there are 345 boys at your school, how many would you expect to be colour blind?

5 The probability of an egg having two yolks is 0.001. If a farm produces 90 dozen eggs each week, how many would you expect to have two yolks?

6 In the game of poker, each player is dealt a hand of five cards. A full house is when a hand contains three of a kind and a pair, e.g. 6, 6, 6, 2, 2. The probability that a hand contains a full house is 0.001441. If 64 hands are dealt during the course of a game, how many of these would be expected to contain a full house?

Probabilities from bar graphs

The graph shows the number of siblings (brothers and sisters) of 40 students.

'Frequency' means the number of students getting each score.

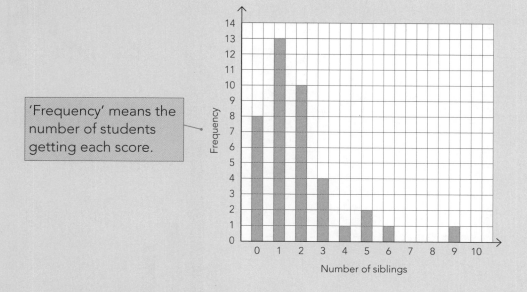

a How many students have two siblings?

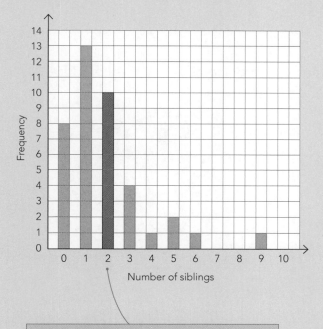

Because the column above 2 goes up to 10, we know that 10 students have two siblings.

b What is the probability that a student has more than three siblings?

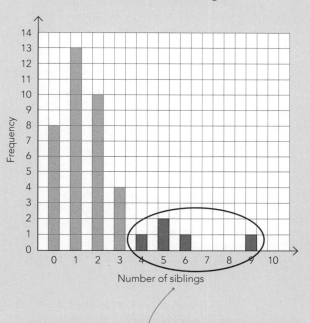

Because there are five students in the columns to the right of 3, and there were 40 students in total, the probability that a student has more than three siblings $= \frac{5}{40} = 0.125$.

Look at the following graphs and answer the questions. Where appropriate, round your answers to 3 sf.

1   Fred recorded the number of goals scored by his team in each of their hockey games throughout the season. His totals are shown on the graph below.

    a   How many games did they play?

    _____

    b   What is the probability that they scored one goal in a game?

    _____

    c   In what percentage of games did they score fewer than three goals in a game?

    _____

    d   How many goals in total did they score during the season?

    _____

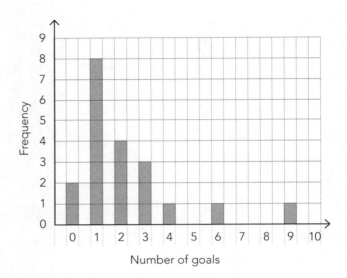

2   A farmer with a small flock of ewes recorded the number of lambs each animal produced. He plotted these on the graph below.

    a   How many ewes were in his flock?

    _____

    b   What percentage of ewes had one lamb?

    _____

    c   What is the probability that a ewe had one or two lambs?

    _____

    d   How many lambs in total were produced by the flock?

    _____

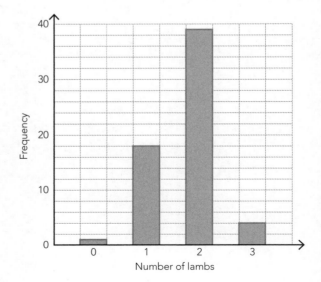

3 Ellie performed an experiment by tossing a six-sided die a large number of times and recording the results. The graph of the distribution is shown below.

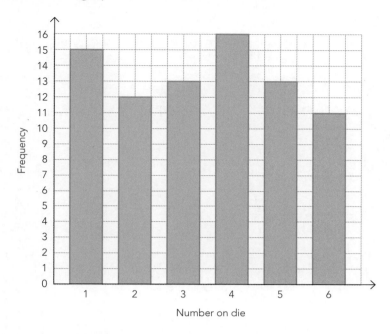

a How many times in total did she toss the die?

_____

b Based on her results, what was her experimental probability of getting a four?

_____

c What is the theoretical probability of getting a four when a fair die is tossed?

_____

d Explain why your answers in **b** and **c** might be different.

_____

e If she tossed a fair die 120 times, how many sixes would you expect her to get?

_____

f If she performed her experiment again, what would you expect her graph to look like?

_____

_____

g She set her computer to do a simulation for 1000 tosses of a die. How would you expect the graph of simulated results to differ from the graph of her experiment?

_____

_____

Probabilities with two events

- We sometimes need to calculate the probability of two events occurring.
- It usually makes calculations easier if you create a **table** showing all the possible outcomes.

1 With equally likely outcomes

- Where two events occur and each event has equally likely outcomes, you can calculate the probabilities of combined events from a two-way table.

Example: Noa tosses a coin and throws a die. The table shows the possible outcomes.

First event: tossing a **coin**.

Second event: throwing a **die**.

	1	2	3	4	5	6
H	H 1	H 2	H 3	H 4	H 5	H 6
T	T 1	T 2	T 3	T 4	T 5	T 6

$P(T) = \dfrac{6}{12}$ or 0.5

$P(6) = \dfrac{2}{12}$ or $0.1\dot{6}$

$P(T \text{ and a } 6) = \dfrac{1}{12}$ or $0.08\dot{3}$

$P(T \text{ or a } 6) = \dfrac{7}{12}$ or $0.58\dot{3}$

There are **12** different but equally likely outcomes.

1 Frank is playing a board game in which two dice are tossed at the same time. One is **black** and the other is **green**. The possible outcomes are shown in the table below.

a Complete the table.

	1	2	3	4	5	6
1	1 1	1 2				
2	2 1	2 2				
3						
4						
5						
6						

b How many different possible outcomes are there? _____

c List the outcomes that produce a total of 11. _____

d P(total is 11) = _____

e P(total is 7) = _____

f P(total is 7 or 11) = _____

g P(total is less than 4) = _____

h P(total is 4 or more) = _____

i P(total is even) = _____

j P(double) = _____

k P(green die is more than black die) = _____

ISBN: 9780170477437

2 In another game, Henare uses a spinner with the numbers **1**, **2**, **4** and **8**, and he tosses a die.

a Complete the table to show all the possible outcomes.

	1					6
1		1 2				
	2 1			2 4		
			4 3			
8						8 6

b P(total is 2) = _____   **c** P(double) = _____

d P(total is odd) = _____   **e** P(total is more than 10) = _____

3 Aria has a coin and an octahedral die (eight equal faces). The die has the numbers 1 to 8 on its faces.

a Use the space below to draw a table of the possible outcomes.

b P(8) = _____   **c** P(T and 7) = _____

d P(T or 7) = _____   **e** P(H and an odd number) = _____

2 When probabilities of events are known

- Where **two events** occur and you **know their probabilities**, you can calculate the probabilities of combined events in two ways.

1 Addition rule: if one event **OR** another happens $\Rightarrow$ **ADD** the probabilities.

2 Multiplication rule: if one event **AND** another happens $\Rightarrow$ **MULTIPLY** the probabilities.

Example: A bag contains 20 lollies: 6 are lime flavoured, 9 are raspberry, 4 are lemon and 1 is blackberry.

$$\text{P(lime or raspberry)} = \frac{6}{20} + \frac{9}{20}$$

$$= \frac{15}{20} \text{ or } \frac{3}{4} \text{ or } 0.75$$

$$\text{P(lime or raspberry or lemon)} = \frac{6}{20} + \frac{9}{20} + \frac{4}{20}$$

$$= \frac{19}{20} \text{ or } 0.95$$

If lollies are picked out and then **replaced** in the bag:

$$\text{P(lime and raspberry)} = \frac{6}{20} \times \frac{9}{20}$$

There are still 20 lollies left.

$$= \frac{54}{400} \text{ or } \frac{27}{200} \text{ or } 0.135$$

If the lollies are picked out and then **not replaced** (eaten), the probabilities change.

$$\text{P(lime and raspberry)} = \frac{6}{20} \times \frac{9}{19}$$

$$= \frac{54}{380} \text{ or } 0.142$$

There are only 19 lollies left.

$$\text{P(two limes)} = \frac{6}{20} \times \frac{5}{19}$$

There are only 5 lime lollies left.

$$= \frac{30}{380} \text{ or } 0.079$$

1 Calculate the following probabilities using the same bag of lollies. Where needed, round your answers to 2 dp.

a P(lime or lemon) = _____

b P(lime or lemon or blackberry) = _____

_____

_____

Lollies are picked out of the bag, their flavour recorded, and then replaced. Then another is picked out.

c P(lemon and lime) = _____

d P(two raspberries) = _____

_____

_____

Lollies are picked out of the bag and then eaten. Then another is picked out.

e P(lemon and blackberry) = _____

f P(two lemons) = _____

_____

_____

g P(three raspberries) = _____

h P(two blackberries) = _____

_____

_____

3 Using probability trees

- Probability trees are very useful for calculating probabilities where several events occur.

Example: Sala had a bag of marbles. It contained 3 blue ones and 7 green ones. Without looking, she selected a marble and put it aside. She then selects a second marble.

a What is the probability that she picked two blue marbles?

b What is the probability that she picked one of each colour?

Steps:

1 Decide what the **events** are, and their order:

Event 1 will be the colour of the first marble.

Event 2 will be the colour of the second marble.

Write the **events** at the **ends** of the branches:

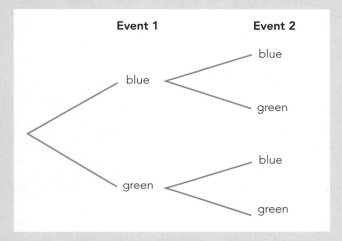

2 Add the **probabilities** of each event to the **middle** of each branch:

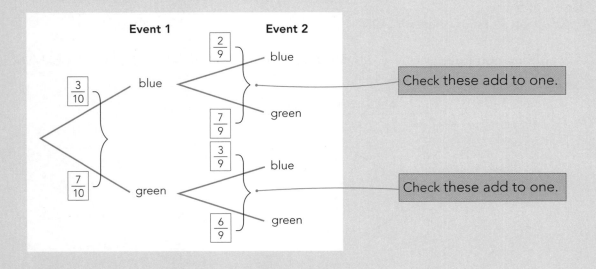

3 **Check** that the probabilities for every branch add to one.

4 **List** the outcomes at the ends of each branch, and calculate the probabilities at each end. You **multiply** the probabilities along each branch because Event 1 **and** Event 2 must occur.

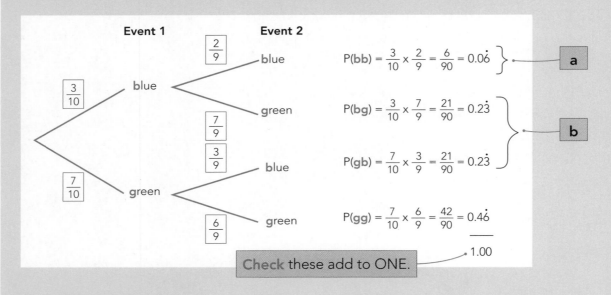

5 **Check** that your probabilities in the right-hand column **add** to **one**. This is because **one** of these options **must** occur.

6 In the right-hand column, highlight the event(s) required, along with their probabilities. **Add** these to find the overall probability required.

a P(bb) = 0.0$\dot{6}$
b P(bg or gb) = 0.2$\dot{3}$ + 0.2$\dot{3}$ = 0.4$\dot{6}$

Useful tips:

1 **Addition Rule:** if one event **OR** another happens ⟹ **ADD** the probabilities.

2 **Multiplication Rule:** if one event **AND** another happens ⟹ **MULTIPLY** the probabilities.

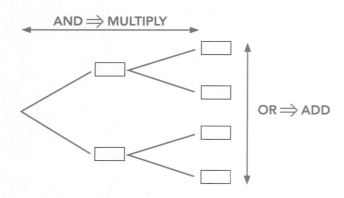

Use probability trees to answer the following questions.

1 A game at the local fair will give you a prize if you roll a four on a 'fair' die (dice). The probability tree below shows the possible outcomes for the first two throws of a die.

a Complete the probability tree.

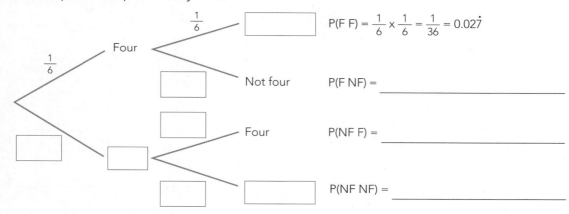

$P(F\ F) = \frac{1}{6} \times \frac{1}{6} = \frac{1}{36} = 0.02\dot{7}$

P(F NF) = _____

P(NF F) = _____

P(NF NF) = _____

b What is the probability that a player gets a four on the first throw?

_____

c What is the probability that a player gets a four and then 'not four'?

_____

d What is the probability that a player rolls two fours in the first two throws?

_____

e What is the probability that a player gets exactly one four in the first two throws?

_____

f What is the probability that a player doesn't get a four in the first two throws?

_____

g What is the probability that a player gets at least one four in the first two throws?

_____

h What do you notice about your last two answers? Why is this?

_____

2   Sue had a bag of marbles. It contained 7 yellow ones and 4 red ones. Without looking, she selected a marble and put it aside. She then selects a second marble.

   a   Complete the probability tree.

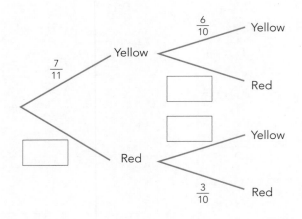

$$P(YY) = \frac{7}{11} \times \frac{6}{10} = \frac{42}{110} = 0.38\dot{1}$$

P(YR) = _____

P(RY) = _____

P(RR) = _____

   b   What is the probability that she selects two yellow marbles?

_____

   c   What is the probability that she selects one of each colour?

_____

   d   What is the probability that she selects two of the same colour?

_____

   e   What is the probability that she doesn't select a yellow marble?

_____

   f   What is the probability that she selects at least one red marble?

_____

   g   What is the probability that after selecting two marbles, there are exactly two red marbles left in the bag?

_____

   h   If Sue repeated this experiment 200 times, how many times could she expect to get two red marbles?

_____

4 Using two-way tables

- When there is data on two different characteristics for each member of a population, you can display these in a two-way table.

Example: At a school of 714 students, each is asked about their use of Snapchat and Instagram. The results are shown in the table below.

	Use Instagram	Don't use Instagram
Use Snapchat	382	118
Don't use Snapchat	46	168

It is a good idea to add totals to your table:

	Use Instagram	Don't use Instagram	Totals
Use Snapchat	382	118	500
Don't use Snapchat	46	168	214
Totals	428	286	714

Check that both add to **714**.

a What is the probability that a student uses Snapchat?

$$P(\text{Snapchat}) = \frac{500}{714} = 0.7000$$

b What is the probability that a student uses neither Snapchat nor Instagram?

$$P(\text{neither}) = \frac{168}{714} = 0.235$$

c If two students from this school are selected at random, what is the probability that both students use Snapchat?

One student **and** the other use Snapchat so you **multiply** their probabilities.

$$P(\textbf{both} \text{ students use Snapchat}) = \frac{500}{714} \times \frac{499}{713}$$
$$= 0.490$$

d If an Instagram user is selected at random, what is the probability that they also use Snapchat?

$$P(\text{Instagram user uses Snapchat}) = \frac{382}{428} = 0.893$$

We are just concerned with the 428 Instagram users.

e Compare the answers to **a** and **d**. What do they tell you about users of Instagram and Snapchat at this school?

Students are more likely to use Snapchat (P = 0.893) if they are also an Instagram user. The probability of using Snapchat in the group as a whole was only 0.7.

f At another similar school with a roll of 528 students, how many students would you expect to use both Snapchat and Instagram?

$$\text{Expected number of users of both} = \frac{382}{714} \times 528$$
$$= 282.49$$
$$= 282 \text{ or } 283 \text{ students}$$

Use the data in the following tables to answer the questions. Where needed, round your answers to 2 dp.

1 Data was collected from the same school about how many Year 12 and 13 students went to the Formal. Complete the table.

	Year 12	Year 13	Totals
Went to the Formal	84	96	
Didn't go to the Formal	30	6	
Totals			

a What percentage of students went to the Formal? _____

b What is the probability that a randomly selected student was in Year 12 and went to the formal?

_____

c What is the probability that a Year 12 student went to the Formal?

_____

d If two students from Years 12 and 13 were selected at random, what is the probability that both went to the Formal?

_____

2 Theo surveyed 96 students. He asked them whether or not they had ever been snowboarding. Complete the table.

	Been snowboarding	Had not been snowboarding	Totals
Year 11	12		
Year 9		36	
Totals	21		

a What is the probability that a randomly selected student had not been snowboarding?

_____

b What is the probability that a randomly selected student was Year 11 who had been snowboarding?

_____

c What is the probability that a randomly selected Year 9 had been snowboarding?

_____

d What is the probability that two randomly selected students from his survey had both been snowboarding?

_____

e The school has 443 Year 11 students. How many would you expect have not been snowboarding?

_____

3 The table below shows the number of passengers on a flight to Sydney, along with their meal preference and fare type.

	First Class	Business Class	Economy	Totals
Noodles	7			
Salad		12		82
Totals	18		245	283

a What was the probability that a passenger ordered salad?

_____

b What was the probability that a passenger in First Class ordered noodles?

_____

c What was the probability that a Business Class passenger ordered salad?

_____

d What was the probability that an Economy passenger ordered salad?

_____

e What percentage of passengers were in First or Business Class?

_____

4 A survey was done on the teeth of Year 8 students in two towns. One town had fluoride added to its water supply and the other did not. The survey recorded whether or not there was evidence of tooth decay. Complete the table.

	Fluoride	No fluoride	Totals
No tooth decay		196	
Tooth decay	154		274
Totals		316	828

a What is the probability that a student had no tooth decay?

_____

b What is the probability that a student living in an area without fluoride had no tooth decay?

_____

c What is the probability that a student living in an area with fluoride had no tooth decay?

_____

d What is the probability that two randomly selected Year 8 students from these areas both had tooth decay?

_____

e There are approximately 66 700 Year 8 students in New Zealand. How many of these would you expect to have tooth decay?

_____

Probability investigations

1 What's the point? Explore a purpose

- Briefly describe your investigation.
- Write a **purpose** or introduction that includes the investigative question or statement.
- Describe any benefits or the usefulness of your investigation.

Examples:

1 Huia is planning a game of chance for the school fair. She has created six cards. Two show a ruru (morepork) and four show a manaia (seahorse).

She will:
- turn the cards face down
- ask the player to pick up one card and keep it in their hand
- then pick a second card.

She will give a prize to the player if they have picked one of each design.

In order to decide the value of the prize, she needs to know the probability of a player picking two cards with the same design.

2 A class of 30 are trying to improve their memories using Kim's game. Ten random but familiar objects are placed on a tray, and the players have one minute to memorise the objects. The tray is removed and each student has to list the names of as many objects as they can within the next five minutes.

Their teacher would like to know the probability that a student in the class can remember 8 or more of the objects.

ISBN: 9780170477437

2 What do we need to think about? Plan your investigation

You need to:

a List all the possible **outcomes** and the **event** you are interested in. These must relate to the question/purpose.

b Describe **how** you will carry out your investigation.

c Describe at least two sources of **variation**, and how you will control these.

d State the **number** of trials you will perform, or the **number** of observations you will make. This must be at least 30.

a Trials, outcomes and events

- A **trial** is one run through of your experiment.
- The **outcomes** of an experiment are all the possible results from one trial. This is also known as the **sample space**.
- An **event** is the outcome/outcomes from a trial which is/are of interest in your question.

Examples:

1 Huia's game

One trial	Sample space	Event
Drawing two cards.	• *Drawing two cards the same.* • *Drawing different cards.*	*Drawing different cards.*

> Make sure that every possibility is covered by your two outcomes. In this case, the player **must** draw 0, 1 or 2 ruru.

2 Kim's game

One trial	Sample space	Event
A student plays the game once.	• *Student remembers between 0 and 7 items.* • *Student remembers 8, 9 or 10 items.*	*Student remembers 8, 9 or 10 items.*

1 Complete the table.

Question/statement	One trial	Sample space	Event
Lollies ●●●●●●●●●●●● Liz has a bag containing 12 lollies — 7 black and 5 green. If she selects and gives the lolly to a classmate, then selects a second, what is the probability she selected one of each colour?	_____ _____ _____	• Selecting two lollies of the same colour. • _____ _____	_____ _____ _____
Paper planes ✈ Tane's class is making paper planes and each class member will have three throws of their plane in a final competition. Tane would like to know the probability that his plane goes at least 5 m before hitting the ground in at least two of his throws.	_____ _____ _____ _____ _____	• _____ _____ _____ • _____ _____	_____

b Describe how you will carry out your investigation

- You will need to describe **how** you will **perform** your investigation in enough detail that somebody else could copy it exactly.
- Describe how the data will be **collected** and **recorded**.

Examples:

1 Huia's game

Huia will turn all six cards so they are face down. She will draw two cards without replacing the first one. For each trial she will record the initials of the cards drawn: R for ruru and M for manaia. She will shuffle the cards in between each trial.

2 Kim's game

The teacher should spread the 10 familiar objects on a tray. The students should get one minute to memorise the objects and then five minutes to list the names of the objects.

2 Complete the table.

Question/statement	Description of steps and recording of results
Lollies ●●●●●●●○○●●○ Liz has a bag containing 12 lollies — 7 black and 5 green. If she selects a lolly and gives the lolly to a classmate, and then selects a second, what is the probability she selected two of the same colour?	
Paper planes Tane's class are making paper planes and each class member will have three throws of their plane in a final competition. Tane would like to know the probability that his plane goes at least 5 m before hitting the ground in at least two of his throws.	

   ISBN: 9780170477437

c Sources of variation

- Sources of **variation** are things that might **change the probability** of success during your experiment.

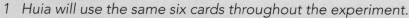

Examples:

1 Huia's game

1 *Huia will use the same six cards throughout the experiment.*
2 *She will regularly check that none of the cards have become marked on their backs. This is to ensure that all six cards appear identical.*
3 *She will make sure the cards are thoroughly shuffled after each trial. Otherwise some cards might get picked more often than others because of their position. She will make sure that there is no possibility that the player can see what is on the cards (e.g. don't use a glass table).*
4 *The results are recorded accurately.*

2 Kim's game

1 *The objects should be spread out so they are all easily seen.*
2 *The objects should all be familiar objects so students will know their names.*
3 *The teacher should just test a small group of students at one time, so that they all get a good view of the tray.*
4 *The teacher should make sure that no students get to see the objects on the tray before the experiment is performed.*
5 *The teacher should ensure that every student gets exactly one minute to look at the tray and five minutes in which to write their lists, and that they don't cheat.*

3 Complete the table.

Question/statement	Sources of variation
Lollies ●●●●●●●○○○○○ Liz has a bag containing 12 lollies — 7 black and 5 green. If she selects a lolly and gives the lolly to a classmate, and then selects a second, what is the probability she selected two of the same colour?	
Paper planes Tane's class are making paper planes and each class member will have three throws of their plane in a final competition. Tane would like to know the probability that his plane goes at least 5 m before hitting the ground in at least two of his throws.	

d Number of trials

Examples:

1 Huia's game

I will do 50 trials because this will be quite quick to do, and I think 50 trials should give me a good estimate of the probability.

2 Kim's game

The teacher will test all 30 members of the class.

4 Complete the table and discuss your answers with your class:

Question/statement	Number of trials
Lollies ●●●●●●●●●●●● Liz has a bag containing 12 lollies — 7 black and 5 green. If she selects a lolly and gives the lolly to a classmate, and then selects a second, what is the probability she selected two of the same colour?	
Paper planes Tane's class are making paper planes and each class member will have three throws of their plane in a final competition. Tane would like to know the probability that his plane goes at least 5 m before hitting the ground in at least two of his throws.	

At this point, your teacher will check that your purpose is suitable.

3 Record and display your results

You need to:
- **List** your results as you collect them.
- Create a **table** to show tallies of your results and probabilities.
- Draw a **frequency** and/or a **probability graph** to show your results.

1 Huia's game

List of results. These show the initials of the animals in each pair.

MR RR MM MM MR	MM MM RM MM RR
RM MM MM MR MR	MR MR MM RM RM
MM MM MR MR RM	MM MM MM MR RM
RR MM MM MM MR	RM MM RM MM MR
RM MR MM MM MM	RR RM MM MM RM

It is useful to record your results in groups of **five**, particularly if you want to plot the probabilities throughout the experiment.

A table showing results — you must do this.

	Manaia	Ruru	Totals
Manaia	23	12	35
Ruru	11	4	15
Totals	34	16	50

Before you continue, check that your **frequencies add to the number of trials.**

A table showing probabilities — this is useful.

	Manaia	Ruru	Totals
Manaia	$\frac{23}{50} = 0.46$	$\frac{12}{50} = 0.24$	$\frac{35}{50} = 0.7$
Ruru	$\frac{11}{50} = 0.22$	$\frac{4}{50} = 0.08$	$\frac{15}{50} = 0.3$
Totals	$\frac{34}{50} = 0.68$	$\frac{16}{50} = 0.32$	1

Check that your **probabilities add to 1.**

You must draw a frequency and/or a probability graph to show your results.

Frequency graph

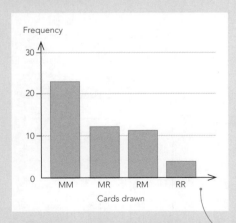

Probability graph

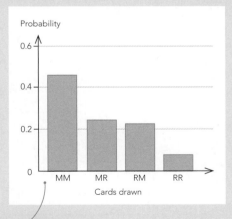

Notice that these graphs are the same shape, but have different vertical scales.

2 Kim's game

List of results. These show the number of items remembered by each student.

Results

List of results (number of items remembered)	
7 5 3 6 9	10 6 7 7 5
4 6 7 7 5	4 7 6 6 6
8 6 7 5 9	6 8 8 7 6

Tables to show the frequencies and probabilities of getting each score

Frequencies

Score	0	1	2	3	4	5	6	7	8	9	10	Total
Number of students	0	0	0	1	2	4	9	8	3	2	1	30

Probabilities

Score	3	4	5	6	7	8	9	10	Total
Probabilities	$\frac{1}{30} = 0.0\dot{3}$	$\frac{2}{30} = 0.0\dot{6}$	$\frac{4}{30} = 0.1\dot{3}$	$\frac{9}{30} = 0.3$	$\frac{8}{30} = 0.2\dot{6}$	$\frac{3}{30} = 0.1$	$\frac{2}{30} = 0.0\dot{6}$	$\frac{1}{30} = 0.0\dot{3}$	1

Graphs to show the frequencies and probabilities of getting each score

Frequency

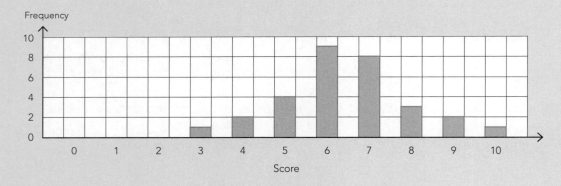

Probability

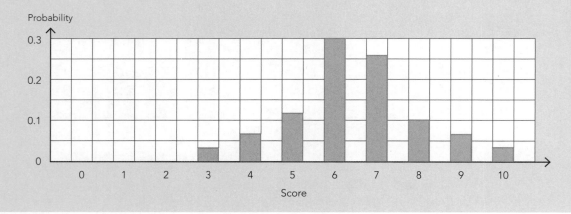

 ISBN: 9780170477437

Answer the following.

1 **Lollies.** If Liz selects a lolly and gives the lolly to a classmate, and then selects a second, what is the probability she selected two of the same colour?

Results

List of results (colours of first and second lolly)	
BB GB GB BG GG	GG BB GB BB GG
GG GB BG BG BG	GB BG BG BB BB
GB BG BG BB BB	BG GG BB BB BB

a Complete the tables to show the frequencies and probabilities of getting each sum.

Frequencies

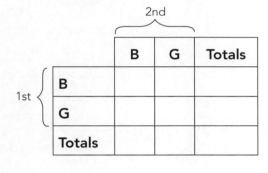

2nd

1st		B	G	Totals
	B			
	G			
	Totals			

Probabilities

	B	G	Totals
B			
G			
Totals			1

b Draw graphs.

Frequency

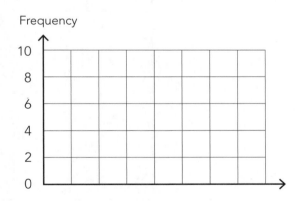

Frequency

10
8
6
4
2
0

Probability

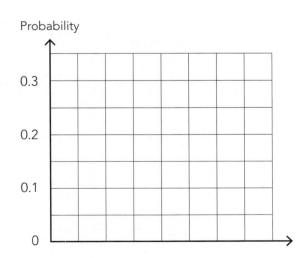

Probability

0.3

0.2

0.1

0

2 **Paper planes.** Tane would like to know the probability that his plane goes at least 5 m before hitting the ground in at least two of his three throws.

Results

List of results (number of throws going at least 5 m)	
3 3 1 2 2	1 0 1 2 2
0 2 2 2 3	2 2 1 2 3
2 1 1 3 1	3 0 1 2 3

a Complete the tables to show the frequencies and probabilities of getting each sum

Frequencies

	0	1	2	3	Total
Number that travelled 5 m					30

Probabilities

	0	1	2	3	Total
Probability					1

b Draw graphs.

Frequency

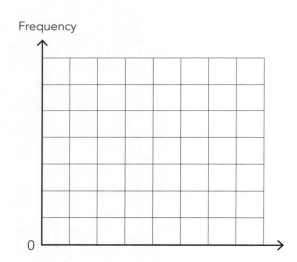

Probability

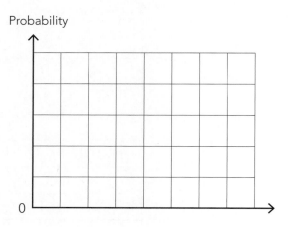

4 Describe what you see in your results and calculate the experimental probability

You need to:
- **Describe** what you see in your frequency table.
- **Describe** what you see in your probability table.
- **Calculate** the experimental results.

1 Huia's game

Frequencies

	Manaia	Ruru	Totals
Manaia	23	12	35
Ruru	11	4	15
Totals	34	16	50

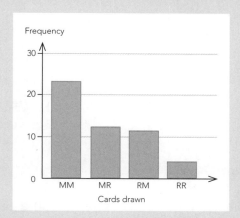

I **notice** that 23 players drew two manaia and only four players drew two ruru.

This **means** most players ($\frac{27}{50}$) drew two cards the same.

Probabilities

	Manaia	Ruru	Totals
Manaia	$\frac{23}{50} = 0.46$	$\frac{12}{50} = 0.24$	$\frac{35}{50} = 0.7$
Ruru	$\frac{11}{50} = 0.22$	$\frac{4}{50} = 0.08$	$\frac{15}{50} = 0.3$
Totals	$\frac{34}{50} = 0.68$	$\frac{16}{50} = 0.32$	1

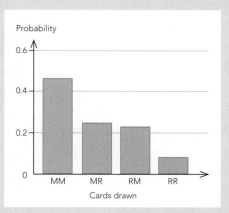

I **notice** that the probability that a person drew two ruru is only $0.0\dot{8}$.
This **means** that it is likely that fewer than 10% of players of the game at the fair will get two ruru.

Calculate the experimental probability.

The **experimental probability** that a player drew one of each card (MR or RM)

$$= P(MR) + P(RM)$$
$$= 0.24 + 0.22$$
$$= 0.46$$

2 Kim's game

Frequencies

Score	0	1	2	3	4	5	6	7	8	9	10	Total
Number of students	0	0	0	1	2	4	9	8	3	2	1	30

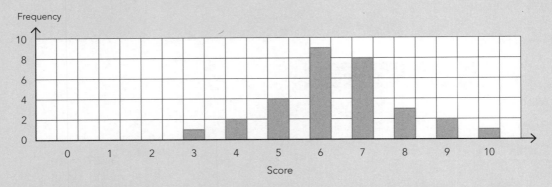

I **notice** that the tallest bars on the graph are for scores of 6 (nine students) or 7 (eight students).
This **means** 17 out of the 30 in the class could remember 6 or 7 of the objects.

Probabilities

Score	3	4	5	6	7	8	9	10	Total
Probabilities	$\frac{1}{30}=0.0\dot{3}$	$\frac{2}{30}=0.0\dot{6}$	$\frac{4}{30}=0.1\dot{3}$	$\frac{9}{30}=0.3$	$\frac{8}{30}=0.2\dot{6}$	$\frac{3}{30}=0.1$	$\frac{2}{30}=0.0\dot{6}$	$\frac{1}{30}=0.0\dot{3}$	1

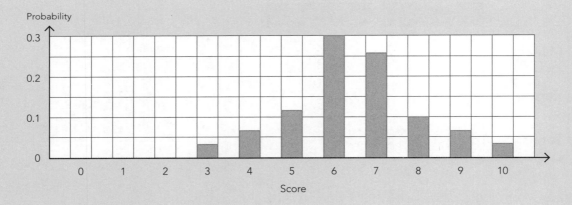

I **notice** that the probabilities for those that scored fewer than 6 are low ($0.0\dot{3}$, $0.0\dot{6}$ and $0.1\dot{3}$).
This **means** that less than a quarter of the class scored fewer than 6 (P = $0.2\dot{3}$).

Calculate the experimental probability.

The **experimental probability** that a score was 8 or more

$$= P(8) + P(9) + P(10)$$
$$= 0.1 + 0.0\dot{6} + 0.0\dot{3}$$
$$= 0.2$$

Describe what you see in your **frequency and probability graphs**.

1   **Lollies.** Use your graphs from page 139.

Frequencies

I notice _____

_____

This means _____

_____

Probabilities

I notice _____

_____

This means _____

_____

The **experimental probability** that _____

$$= P(\underline{\hspace{1.5cm}}) + P(\underline{\hspace{1.5cm}})$$

$$= \underline{\hspace{1.5cm}} + \underline{\hspace{1.5cm}}$$

$$= \underline{\hspace{1.5cm}}$$

2   **Paper planes.** Use your graphs from page 140.

_____

_____

_____

_____

_____

_____

_____

_____

_____

_____

_____

_____

5 Writing a conclusion

You need to:

Answer the question

- Does your experimental probability seem **reasonable**?

Improvements

- How could you have improved the reliability of your results?
- Are there other factors that you haven't mentioned so far that might have affected your results?
- What are the next steps? What could this investigation lead to?

Template:

The experimental probability that _____ is _____

My experimental probability seems **reasonable/surprising/unreasonable** because _____

_____

I could improve the reliability by _____

_____

Other factors that might have affected my results are _____

_____

Next steps: _____

_____

1 Huia's game

The experimental probability that I drew one of each (MR or RM) is 0.24 + 0.22 = 0.46.

This seems reasonable because most of the cards had manaia, so I drew two manaia almost half the time (P = 0.46) and I was unlikely to get two ruru (P = 0.08).

I think my estimate for the probability is fairly reliable because I did 50 trials.

I could improve its reliability by doing more trials.

I don't think that there are any unmentioned factors that might affect my results.

I could use the same method to investigate how the probabilities would change if I altered the number of each card.

2 Kim's game

The experimental probability that a student scored 8 or more was 0.2.

This seems reasonable because I think remembering 8 or more objects would be quite difficult, and only a fifth of the class were able to do it.

I think my estimate for the probability is fairly reliable because I did 30 trials. I could improve its reliability by doing more trials.

The time of day at which the testing was done and what had happened before the mathematics class might have affected the results.

It would be interesting for the teacher to play this every week to see if the students' memories improved.

1 Lollies  31

The experimental probability that _____ is _____.

My experimental probability seems **reasonable/surprising/unreasonable** because _____

_____

_____

I could improve the reliability by _____

_____

_____

Other factors that might have affected my results are _____

_____

_____

Next steps: _____

_____

_____

2 Paper planes

_____

_____

_____

_____

_____

_____

_____

_____

_____

_____

_____

Optional extras

1 Plotting probabilities throughout an experiment

During a probability experiment, it can be useful to plot the probabilities at regular intervals. This graph willl show you the long-run relative frequencies.
- Watching for when the graph stabilises will indicate whether you have done enough repetitions of the experiment.

You need to:
- **Calculate** the experimental probabilities as the investigation progresses.
- **Plot** these on a line graph.
- Describe and discuss what you see in the **line graph**. How do the probabilities change throughout your experiment?

Huia's game
Here are the results from 50 trials.

Note: You could work out these probabilities for every single trial.

List of results	Total number of trials	Total number of outcomes with a ruru and a manaia	P(a ruru and a manaia)	
MR RR MM MM MR	5	2	$\frac{2}{5}$	0.4
MM MM RM MM RR	10	3	$\frac{3}{10}$	0.3
RM MM MM MR MR	15	6	$\frac{6}{15}$	0.4
MR MR MM RM RM	20	10	$\frac{10}{20}$	0.5
MM MM MR MR RM	25	13	$\frac{13}{25}$	0.52
MM MM MM MR RM	30	15	$\frac{15}{30}$	0.5
RR MM MM MM MR	35	16	$\frac{16}{35}$	0.457
RM MM RM MM MR	40	19	$\frac{19}{40}$	0.475
RM MR MM MM MM	45	21	$\frac{21}{45}$	$0.4\dot{6}$
RR RM MM MM RM	50	23	$\frac{23}{50}$	0.46

Two of the first five trials had a ruru and a manaia.

Three out of the first 10 trials had a ruru and a manaia.

The final probability should be the same as your experimental probability.

Probability

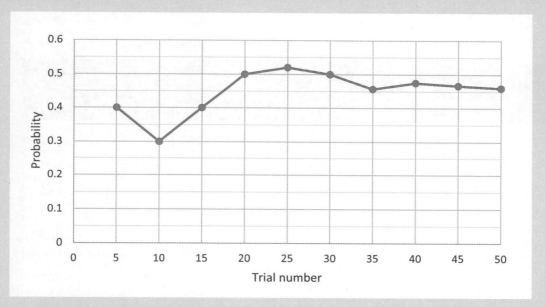

Notice that:
- There is more variation in the first 20 trials, and as the number of trials increases, the variation decreases.
- Towards the end, the graph stabilises at about 0.46.

To be **more certain** of the probability of getting a ruru and a manaia when two cards are drawn, we would need to do **more trials**.

Kim's game

Here are the results from 30 trials.

List of results	Total number of trials	Total number of outcomes where a student in the class scored 8 or more	P(scored 8 or more)	
7 5 3 6 9	5	1	$\frac{1}{5}$	0.2
10 6 7 7 5	10	2	$\frac{2}{10}$	0.2
4 6 7 7 5	15	2	$\frac{2}{15}$	0.1̇3
4 7 6 6 6	20	2	$\frac{2}{20}$	0.1
8 6 7 5 9	25	4	$\frac{4}{25}$	0.16
6 8 8 7 6	30	6	$\frac{6}{30}$	0.2

Probability

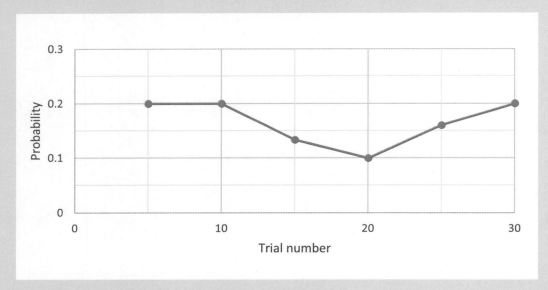

Notice that:

- There is a lot of variation in probability during the 30 trials.
- The graph doesn't stabilise, which suggests that the teacher should try doing it with other equivalent classes in order to do **more trials**.

 ISBN: 9780170477437

Answer the following.

1   Lollies   ●●●●●●●●●●●●●

a   Complete the table in order to show the probability of getting 10 or more:

List of results	Total number of trials	Total number of trials where Liz selected two of the same colour	P(two the same)	
BB GB GB BG BB	5	2	$\frac{2}{5}$	0.4
GG BB GB BB GG	10	6	$\frac{6}{10}$	0.6
GG GB BG BG BG	15			
GB BG BG BB BB	20			
GB BG BG BB BB	25			
BG GG BB BB BB	30			

b   Complete the graph in order to show the probability of getting 10 or more.

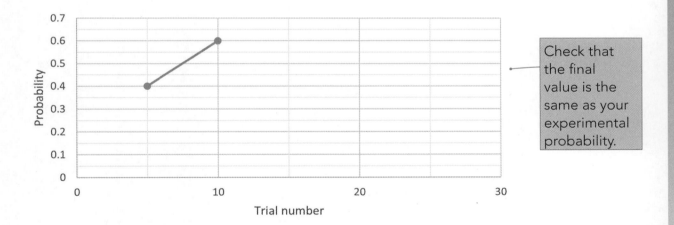

Check that the final value is the same as your experimental probability.

c   Describe the variation in the graph. _____

_____

Towards the end, the graph stabilises _____

To be more certain of the probability of getting two lollies of the same colour, we would

need to _____

2 Paper planes

a Complete the table in order to show the probability that Tane's plane flew further than 5 m in at least two attempts.

List of results	Total number of trials	Total number of times that Tane's plane flew further than 5 m in at least two attempts	P(at least two flew more than 5 m)	
3 3 1 2 2		4		
1 0 1 2 2		6		
0 2 2 2 3				
2 2 1 2 3				
2 1 1 3 1				
3 0 1 2 3				

b Complete the graph in order to show the probability that Tane's plane flew further than 5 m in at least two attempts.

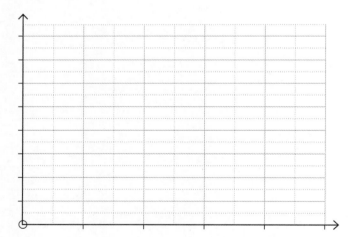

c Discuss what these results tell you.

_____

_____

_____

_____

_____

2 Calculating the theoretical probability

- For some experiments you **can** calculate the theoretical probability.
- Experiments in which you **can** calculate experimental probabilities are usually those that involving equally likely outcomes (dice, coins, lollies which are identical apart from their colour, etc).
- Experiments in which you **cannot** calculate experimental probabilities tend be those involving skills (such as kicking a rugby ball), natural events (such as thunderstorms), etc.

You should:
- **Show** how you calculated the experimental probability.
- **Compare** your experimental probability with the theoretical probability.
- **Discuss**, including why they might be similar or different.

Huia's game

With the first card: the probability of getting a ruru is $\frac{2}{6}$, and the probability of getting a manaia is $\frac{4}{6}$.

That leaves five cards.

If a ruru was drawn first, then there is only one ruru left, so $P(RR) = \frac{2}{6} \times \frac{1}{5} = \frac{2}{30} = 0.0\dot{6}$.

If a manaia was drawn first, then there are only three manaia left, so $P(MM) = \frac{4}{6} \times \frac{3}{5} = \frac{12}{30} = 0.4$.

So the theoretical probability of drawing two cards the same $= 0.0\dot{6} + 0.4 = 0.4\dot{6}$.

Comparison with the experimental probability and discussion
The experimental probability was 0.46, which very close to $0.4\dot{6}$. I did 50 trials which is a reasonable number. However, I would expect the experimental probability would have been even closer to $0.4\dot{6}$ if I had done more trials.

Kim's game
This is an example of an experiment for which the theoretical probability **cannot** be calculated because we do not know the probability that a student will remember 8 or more items.

Calculate theoretical probabilities for the following situations and compare each with the given experimental probabilities.

1   Lollies

Calculation of the theoretical probability:

_____

_____

_____

_____

_____

_____

_____

Comparison with the experimental probability and discussion:

_____

_____

_____

_____

2   Paper planes

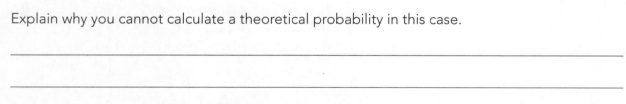

Explain why you cannot calculate a theoretical probability in this case.

_____

_____

_____

_____

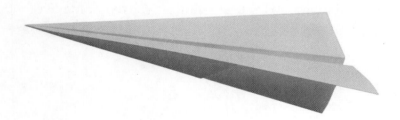

ISBN: 9780170477437

Putting it all together

You will need to do the following:

1 Introduction
- Briefly **describe** the context of your investigation.
- Write or identify the investigative **question** or statement.
- Describe a **purpose** for your investigation.

2 Plan your investigation
a List all the possible **outcomes** (sample space) and the **event** you are interested in. These must relate to the question/purpose.
b Describe **how** you will carry out your investigation.
c Describe at least two sources of **variation**, and how you will control these.
d State the **number** of trials you will perform, or the **number** of observations you will make. This must be at least 30.

3 Variation
- Describe at least two types of **variation**, and how you will manage these.

4 Record and display your results
- **List** your results as you collect them.
- Create a **table** to show tallies of your results and probabilities.
- Draw a **frequency** and/or a **probability graph** to show your results.

5 Describe what you see in your results and calculate the experimental probability
- Describe what you see in your **frequency** graph.
- Describe what you see in your **probability** graph.
- Calculate the **experimental** results.

6 Write a conclusion
- **Answer the question**
  — Does your experimental probability seem **reasonable**?
- **Reflections**
  — How could you have improved the reliability of your results?
  — Are there other factors that you haven't mentioned so far that might have affected your results?
  — What are the next steps? What could this investigation lead to?

Optional extras:

1 Plot the probabilities (long-run frequencies) throughout the experiment.
- **Calculate** the experimental probabilities as the investigation progresses.
- **Plot** these on a line graph.
- **Describe** and **discuss** what you see in the line graph. How do the probabilities change throughout your experiment?

2 Calculate the theoretical probability (not always possible).
- **Show** how you calculated the experimental probability.
- **Compare** your experimental probability with the theoretical probability.
- **Discuss**, including why they might be similar or different.

Annotated example

A primary class has been given 30 packets, each containing three cabbage seeds, as part of a promotion. The publicity states that the seeds are twice as likely to be for green cabbages as red cabbages. Each child planted and labelled their groups of three seeds in the school garden. I would like to know the probability that a child grows at least one of each colour.

For each trial (growing three cabbages) there will be two possible outcomes:
1   A child gets three cabbages the same colour, or
2   A child gets two of one colour and one of the other.

The event I am is interested in is getting two of one colour and one of the other.

Sources of variation:
*   Induced variation — we must assume that all three seeds germinate and that the seeds are all exposed to the same conditions (e.g. soil type, light, water).
*   Situational variation — we assume that the red and green cabbage seeds have been processed similarly and so they are equally likely to germinate.
*   Measurement variation — assume the identification and recording of the cabbage colours are correct.

I should record the number of cabbages the same colour: 3 or 2.

There are 30 trials, one for each student.

List of results	Total number of trials	Total number of trials with two cabbages the same	P(two the same)	
32322	5	3	$\frac{3}{5}$	0.6
23222	10	7	$\frac{7}{10}$	0.7
23322	15	10	$\frac{10}{15}$	0.6̇
22222	20	15	$\frac{15}{20}$	0.75
33322	25	17	$\frac{17}{25}$	0.68
23222	30	21	$\frac{21}{30}$	0.7

Frequency

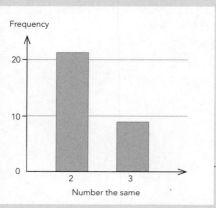

Probability

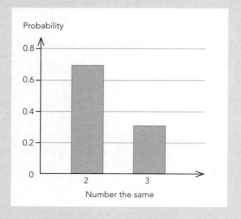

1   Introduction

2   Plan your investigation
Outcomes

Event
Number of trials

3   Variation

4   Display
Recording of results

Table with frequencies and probabilities

Frequency graph
Probability graph

I notice that the bar on the frequency graph for at least one of each colour (2 the same) is more than twice as high as the bar for 3 the same. This means that at least twice as many students grew at least one of each colour as 3 the same.

I notice that the bars in the probability graph show similar proportions. This means that a student was more than twice as likely to grow cabbages of both colours.

Experimental probability of getting at least one of each colour = $\frac{21}{30}$ = 0.7.

Plot of probabilities throughout the experiment:

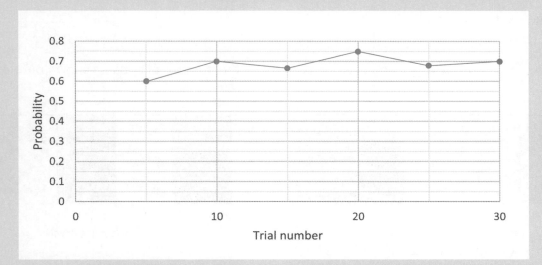

I notice that the graph was variable at the start, but stabilises for the last two values. In order to be sure that it is stabilising and giving a more accurate estimate of the probability, I should do more trials, say 60 (two classes).

Calculation of the theoretical probabilities:

P(getting 3 green cabbages) = $\frac{2}{3} \times \frac{2}{3} \times \frac{2}{3} = \frac{8}{27} = 0.\dot{2}9\dot{6}$

P(getting 3 red cabbages) = $\frac{1}{3} \times \frac{1}{3} \times \frac{1}{3} = \frac{1}{27} = 0.\dot{0}3\dot{7}$

The only other possibility is that a child grows 2 of one colour and 1 of the other:

P(2 of one colour and 1 of the other) = $1 - 0.\dot{2}9\dot{6} - 0.\dot{0}3\dot{7}$

$$= 0.\dot{6}$$

I found that the experimental probability for getting 2 of one colour and 1 of the other was 0.7, which was a little higher than the theoretical probability of 0.$\dot{6}$. I could have improved my estimate by doing more trials. I don't think there are any other factors other than those mentioned earlier which might have affected my results.

Practice task

Monopoly

This activity requires you to write a report about a statistical investigation into a game of chance. You will be assessed on the quality of your discussion and reasoning and how well you link this to the context.

Introduction

Aroha loves playing Monopoly. One side of the square board is shown below.
The rules for starting are:

- Two normal dice are thrown at once.
- You start the game from the square on the right and move clockwise around the board.
- In order to start, you must throw a 6, and then you can move the number of squares shown on the other die.

The object of the game is to buy properties. The properties in the first row are Gore, Edendale, Tumu Tumu Curio Bay, Porpoise Bay and Waipapa Point.

Aroha would like to know the probability that when she throws the dice, she will be able to start and that she will land on a property she can buy.

This task requires you to investigate this probability and write a report for Aroha.

Question and purpose

_____

_____

_____

Plan your investigation

_____

_____

_____

_____

_____

_____

_____

Variation

_____

_____

_____

Record, display and analysis of results

_____

_____

_____

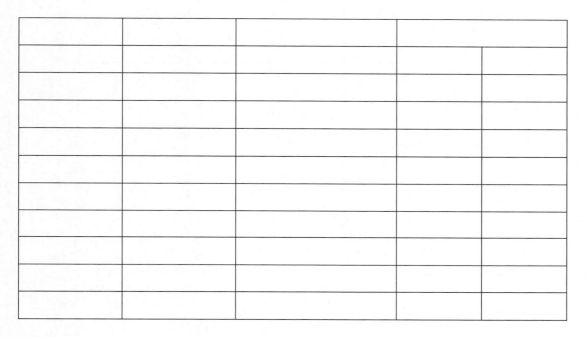

Frequency graph:

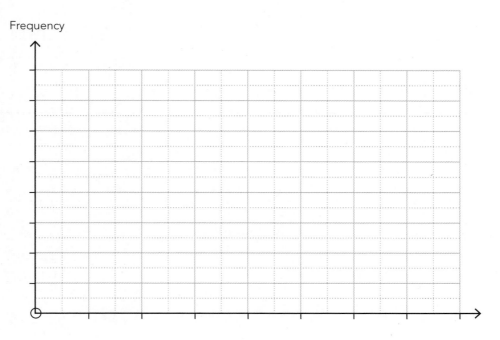

Probability graph:

Probability

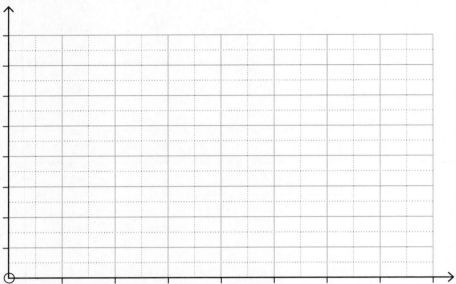

Optional extras

Probabilities throughout the experiment:

Probability

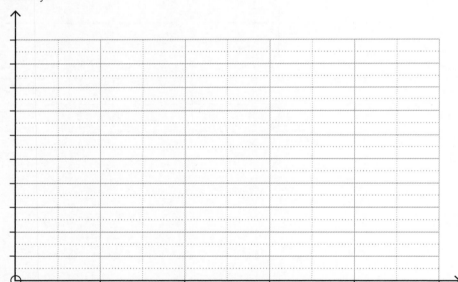

_____

_____

_____

ISBN: 9780170477437

Theoretical probability:

Write a conclusion

_____

_____

_____

_____

Reflections

_____

_____

_____

_____

_____

 Answers

Statistical concepts (pp. 6–7)
Census and sample (p. 6)

	Census or sample	Why?
1	Sample	Too expensive to get feedback from all customers. The information may matter to the company but isn't critical to the New Zealand population.
2	Census	Very important issue that all New Zealanders should get a say in.
3	Sample	Too expensive to get feedback from all customers. The information may matter to the company but isn't critical to the New Zealand population.
4	Census	Very important issue that all New Zealanders should get a say in.
5	Sample	Unless it was a serious consideration, in which case a census would be appropriate.

Types of variables (p. 7)
1 Categorical 2 Continuous
3 Categorical 4 Discrete
5 Continuous 6 Discrete
7 B, C 8 A, E
9 D, F

Investigative questions (p. 9)
1 Summary   Bar graph, pie graph
2 Relationship Scatter plot
3 Summary   Histogram
4 Comparative Box/dot plot
5 Summary   Bar graph, pie graph
6 Comparative Box/dot plot
7 Summary   Histogram
8 Relationship Scatter plot

Sampling and bias (p. 10)
1 a Geckos 7   Geckos 9
   Birds 5   Birds 2
  b The second sample is biased because the numbers in the sample are/are not proportional to the population.
2 c Biased. Samples only students who remain at school during lunchtime and likely to have bought their lunch from the tuckshop.

d Biased. Samples from just one year group.
e Unbiased. Every student in the school has an equal chance of being selected.

Variation (pp. 11–13)
Variation for different types of data (pp. 12–13)
1 Positions of fingers
Device used for measurement
Angle between arms and torso
2

	Type of variation	How could it be managed?
a	Induced	All measurements should be made at the same temperature.
b	Measurement	Have two people timing how long it takes.
c	Situational	All measurements should be made at about the same time of day.

3

	Type of variation	How it could have been managed
a	Measurement	Have two people measuring — one to measure and one to hold the geckos still.
b	Sampling	Enclosing an colony and ensuring that every gecko was caught, not just the big or small ones, males or females.
c	Induced	Handling the geckos as gently as possible and, if possible, marking some geckos to check their survival.

Data check (pp. 14–15)
1 Heights of students in a Year 12 class (cm):
5' 4" and 1082
The measurement 5' 4" is a height measured in feet and inches. This should be converted to centimetres.
The measurement 1082 cm is highly unlikely to be the height of a Year 12 student. Their height should be measured again.
2 Number of hours slept last night: 0 and 29
It seems unlikely that someone didn't sleep at all overnight. This should be double-checked.
It's impossible to sleep 29 hours in one night. This should also be double-checked or removed.
3 The price of motel room for one night: $17
$17 seems far too cheap for a motel room for the night. Double-check this and possibly remove the data.

Data analysis (pp. 15–19)
Measures of centre (pp. 15–17)

1   Mean = 12.5   Median = 12
    Mode = 1
2   Mean = 21.3   Median = 20.5
    Mode = 12 and 27
3   Mean = 2.25   Median = 2.6
    Mode = no mode
4   The mean and median are different.
    There is one large value (93), which affects the mean but not the median.
5   The mean and median are similar.
    The data is symmetrically distributed with no unusual values.
6   **a** 6   **b** 6
7

Set A	3	20	7
Set B	17	7	6
Set C	15	1	14

Measures of spread (pp. 18–19)

1   Minimum = 7   LQ = 18   Median = 28
    UQ = 40   Maximum = 49
    Range = 42   Interquartile range = 22
2   Minimum = 15   LQ = 23   Median = 35
    UQ = 44   Maximum = 51
    Range = 36   Interquartile range = 21
3   Minimum = 2   LQ = 4.5   Median = 9
    UQ = 15.5   Maximum = 20
    Range = 18   Interquartile range = 11

4   9A   Mean = 5.6   Median = 5.5
    Mode = 5   Range = 4
    9B   Mean = 7.05   Median = 7
    Mode = 8 and 10   Range = 9
    Which class did better? 9B
    9B has a higher mean and median than 9A.
    9B does have the larger range, which means their scores were more variable.

Relationships (pp. 20–48)
Variable selection (p. 21)

SVL and mass	The SVL is likely to be related to mass, as the longer their body, the heavier they are likely to be. However, it doesn't take into account how fat they are.
Elevation and SVL	Elevation could be related to mass, as food sources at different elevations could vary, which would influence how large a gecko can grow.

Scatter plots (pp. 22–30)
Understanding scatter plots (pp. 22)

1   **a** (**168**, **8**) This student is **168** cm tall and their bag mass is **8 kg**.
    b (**155**, **4**) This student is **155** cm tall and their bag mass is **4 kg**.
    c 12 kg   **d** 150 cm
    e 164 cm

Describing features of scatter plots (pp. 23–24)

1   negative   2   positive
    strong   weak
    linear   linear
3   negative   4   positive
    moderate   moderate
    linear   linear
5   negative   6   positive
    strong   strong
    non-linear   linear

Describing the meaning of features (pp. 25–30)

1   **I notice** that the relationship between **the temperature (°C)** and **number of ice creams sold** is **positive**.
    This means as the **temperature** increases, the **number of ice creams sold also** tends to **increase**.
2   **I notice** that the relationship between **age** and **running race time for a 5 km race** is **negative**.
    This means as the **age (years)** increases, the **time (min)** tends to **decrease**.
3   I notice that the relationship between the number of NRL games played and annual salary is positive.
    This means as the number of NRL games increases, the annual salary also tends to increase.
4   **I notice** that there **is a moderate** amount of scatter between the points on the graph.
    This means there is a relationship between age and net worth of some celebrities but it isn't strong.
5   **I notice** that there **isn't much** scatter between the points on the graph.
    This means there is a strong relationship between the lowest daily temperature and highest daily temperature in Arthur's Pass.
6   I notice that there is a lot of scatter between the points on the graph.
    This means there is a weak relationship between the distance travelled to school and the time it took to get to school.
7   **I notice** the **trend** appears to be **linear because** I **could** rule a straight line through the middle of the data.
    This means that as the district size increases, the number of public toilets does/~~doesn't~~ tend to increase at a constant rate.
8   **I notice** the **trend** appears to be **non-linear**

because I **couldn't** rule a straight line through the middle of the data.

This means that as the odometer reading increases, the price ~~does~~/doesn't tend to decrease at a constant rate.

9   I notice the trend appears to be non-linear because I couldn't rule a straight line through the middle of the data.

This means that as the average wind gust increases, the maximum wind gust doesn't tend to increase at a constant rate.

Putting it together (pp. 31–33)

1   **There is a** negative, strong, linear relationship between **temperature** and **the number of pies sold.**

Direction: I notice that the relationship between **temperature** and **the number of pies sold** is negative. This means as the **temperature** increases, the **number of pies sold** tends to **decrease.**

Strength: I notice that there **isn't much** scatter between the points on the graph. This means **there is a strong relationship between temperature and the number of pies sold.**

Trend: I notice the trend appears to be **linear** because I **could rule a straight line through the middle of the data.**

This means that as the temperature increases, the number of pies sold tends to decrease at a constant rate.

2   **There is a** positive, strong, linear relationship between the number of students and the number of staff in New Zealand schools.

Direction: I notice that the relationship between the number of students and the number of staff is positive. This means as the number of students increases, the number of staff also tends to increase.

Strength: I notice that there isn't much scatter between the points on the graph. This means there is a strong relationship between the number of students and the number of staff in New Zealand schools.

Trend: I notice the trend appears to be linear because I could rule a straight line through the middle of the data. This means that as the number of students increases, the number of staff tends to increase at a constant rate.

Unusual features (pp. 34–36)

1   I notice that there are two students who spent less than 10 minutes studying but got marks of between 80 and 85%. I also notice that there is one student who studied for 156 minutes and only gained 19% on their assessment.

This means that two students who didn't spend much time studying did well on the test and one spent a lot of time studying and didn't get a high grade.

2   I notice that there is one artwork which is about 425 years old which is worth $450 000 000. This means that it was by far the oldest and by far the most valuable piece of art.

There are also three that were about 270 years old and worth about $2 000 000 each. This means that these works, although similar in age some others, were far more valuable than all except the one above.

Variation in scatter (p. 37)

1   I notice that the scatter for age of football players and number of games played isn't very even. There is more scatter for older players than younger players. This means that there is more variation in the number of games played by the older players and less variation with the younger players.

2   I notice that there is a lot of scatter in zoo visitors when there is a lower amount of rainfall and less scatter when there is more rainfall.

This means that when there very little rainfall, the numbers visiting the zoo vary widely. When it is raining hard, very few people come to the zoo, so there is far less variation.

Line of best fit (pp. 38–39)

1   2

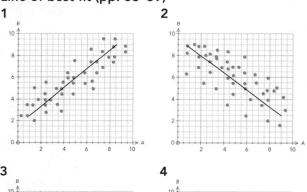

3   4

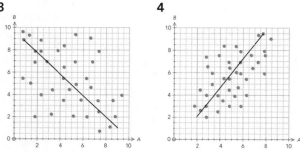

5   6

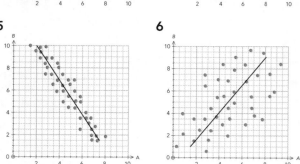

Predictions (pp. 40–42)

For this section, you will need to check your answers with your teacher to make sure they are reasonable.

1 a I predict that if the temperature is 23°C, the number of pies sold will be about **65** but could be anywhere between **55** and **80**.

   b I predict that if the temperature is 16°C, the number of pies sold will be about **130** but could be anywhere between **100** and **170**.

   c I am very/~~reasonably/not very~~ confident in my predictions because the relationship is strong/~~moderate/weak~~.

2 a and b Check with your teacher

   c I am **very** confident in my predictions because the relationship is **strong**.

3 a and b Check with your teacher

   c I am **reasonably confident in my prediction because the relationship is moderate**.

4 a and b Check with your teacher

   c I am **reasonably confident in my prediction because the relationship is moderate**.

Practice task (pp. 46–48)

Question and purpose

I wonder if there is a relationship between the mass of orange-spotted geckos and their snout–vent length (SVL (mm)) from the Otago area.

If we could use the mass of the geckos to estimate their length (SVL), it may reduce the need to handle them and therefore not stress them more than necessary.

Variation

Given that the data was collected for us, we must assume that variation was managed appropriately by the researchers. Here are some potential sources of variation:

- Measurement variation because the geckos were unlikely to have stayed still. We assume the researchers were trained in appropriate gecko-handling techniques to reduce stress and increase accuracy of measurement.
- Situation variation because we don't know how recently the gecko have eaten. This could be managed by measuring the gecko at similar times of the day.

Description

I notice there is strong, positive, linear relationship between gecko mass and SVL.

Direction

I notice that the relationship between mass (g) and SVL (mm) in orange-spotted geckos is positive. This means as the mass of a gecko increases, the SVL tends to increase.

Strength

I notice that there isn't much scatter between the points on the graph. This means that the relationship between mass (g) and SVL (mm) in geckos is strong.

Trend

I notice that the trend appears to be linear because I could rule a straight line through the middle of most of the data. This means that the relationship between mass (g) and SVL (mm) in geckos changes at a constant rate.

Other features

There are three geckos that have more mass and longer SVLs than the rest of the sample. They all have mass at least 17 g and their SVLs are all over 85 mm. Perhaps these are very old geckos.

Prediction

If a gecko had a mass of 9 g, we would expect its SVL to be around 74 mm. However, it could be anywhere between 70 and 78 mm.

Conclusions

There is a strong, positive, linear relationship between mass and SVL of geckos from the Otago region. This means that the larger the mass of the gecko, the longer their snout–vent length is. This result could apply to other geckos in the Otago area but may not be applicable to geckos in other areas or other species.

If another sample was taken, we would likely be measuring different geckos and would therefore get different data. Given the sample size was sufficient, it is likely that we would see a similar trend to the one found above.

To be able to apply our findings to geckos in the Otago area, we are assuming that the sample is representative of this population.

Reflections

- If a larger sample of geckos were taken from the Otago area, there would be more data and it would likely make the relationship clearer. We would be more confident in our conclusions.
- Next steps could be to look into whether there are differences between juvenile and adults or male and female geckos. You could also look at geckos from other parts of New Zealand or different species.

Comparison (pp. 49–81)
Variable selection (pp. 50)

Mass	Adult and Juvenile	It is likely adult geckos are heavier than juvenile. With more data on this, researchers might be able to better estimate geckos as adult or juvenile by weighing them.
Elevation	Jewelled and Orange spotted	It would be interesting to see if different species of gecko are living in different elevations. Different elevations are likely to have different tundra and food availability.

Dot plots (pp. 51–62)
Understanding dot plots (pp. 51–52)
1 a Summary   b 6
   c 13   d 1
2 a 9   b 23
   c All Blacks
      There is more data to the right of the dot plot.
   d Agree
      25 of the players are between 20 and 30 years old. Only 9 are older than 30.

Describing features of dot plots (pp.53–55)
1 a 7   b 7
   c 10
   d There are four pieces of data that are at least 3 units larger than the rest of the data.
2 a 15   b 15
   c 10
   d One student thinks it's appropriate to drive when you are 11 years old. This is at least three years younger than any other age in the data.
3 a The mode for Polar Ice is 8 and that for Nice Cream is 15.
   b The median for Polar Ice will be approximately 8 and the median for Nice Cream will be around 15. This is a difference of 7 pieces of hokey-pokey.
   c The range for Polar Ice is larger (28) than the range for Nice Cream (21).
   d One student found 30 pieces of hokey-pokey in their scoop of Polar Ice, which is far more than anyone else. Another student got only 1 piece of hokey-pokey in their Nice Cream scoop, which is a lot fewer than the rest of the class. Two students got only 6 pieces of hokey-pokey in their Nice Cream scoop, which is also unusual compared with the rest of the class.
4 Left skew   5 Irregular
6 Right skew   7 Bell shaped
8 Bimodal
9 The centre of distribution B is further to the right than that of distribution A.
   Distribution B is more spread out than distribution A.

Distribution A tends towards a **bell shape** and distribution B tends towards a **left skew**.

Comparing dot plots (pp. 57–58)
1 a ✓   b ✗
   c ✗   d ✓
2 a ✗   b ✗
   c ✓   d ✓
3 Pina's coop is slightly skewed to the right with two unusual points at 12 eggs. Gary's coop tends towards a bell-shaped distribution and has no unusual values. The mode and median of Gary's coop are higher than those for Pina's coop. The mode of Gary's is 7, whereas the mode of Pina's is 3. Gary's median is 6 and Pina's is 4 eggs.

Describing meaning of features (pp. 59–62)
1 **Centre:** I notice that the median of set C (**5**) is **smaller** than the median of set D (**7**).
   This means:
   ✗ The sample size of set C is smaller than set D.
   ✗ The data is more spread in data set D than in data set C.
   ✓ The items in set C tend to be smaller than those in set D.
   ✗ The items in set D are all larger than those in set C.
   Spread: I notice that the range of set C (**10**) is **larger** than the range of set D (**8**).
   This means:
   ✗ The sample size for D wasn't large enough.
   ✓ There is more variation in data set C than there is in data set D.
   ✗ There is more variation in data set D than there is in data set C.
   Shape: I notice that data set C tends towards an **irregular** shape.
   This means:
   ✓ If we had more data, we might see a pattern for set C.
   ✓ There is no particular pattern to the distribution.
   I notice that data set D tends towards a **left** skew.
   This means:
   ✓ The data is more spread out on the left hand side.
   ✗ The data in set D is useful and the data in set C is not.
2 **Centre:** I notice that the median of Bernie's bees (**18**) is **larger** than the median of Hemi's hives (**13**).
   This means:
   ✗ Hemi has more bees than Bernie.
   ✓ Bernie's bees tend to produce more honey than Hemi's.
   ✗ Bernie's bees are bigger than Hemi's.
   ✗ All Bernie's bees are better than Hemi's.

Spread: I notice that the range of Bernie' bees (**12**) is **smaller** than the range of Hemi's hives (**15**).

This means:
- ☒ Bernie's bees are better than Hemi's.
- ☑ There is more variation in Bernie's bees than in Hemi's hives.

Shape: I notice that Bernie's bees tends towards a **bell** shape.

This means:
- ☒ No more data should be collected.
- ☒ The hives are producing their maximum amount of honey.
- ☑ The data is symmetrically distributed with most of the data in the middle.

I notice that Hemi's hives tends towards a **right** skew.
- ☒ Most of the data is on the right.
- ☑ The data is more spread out on the right-hand side.

3 **Centre:** I notice that the median length of rāwaru is 42 cm, which is shorter than the median length of mararī, which is 50 cm.

This means the median length of mararī caught in the fishing competition is longer than the median length of rāwaru.

Spread: I notice that range of length for rāwaru is 24 cm, whereas the range of length for mararī is 31 cm.

This means that there is more variation in the length of mararī caught than the rāwaru.

Shape: I notice that the shapes of the distributions for lengths of both fish are irregular but tending towards a normal distribution. This means that most fish caught were of similar sizes, with a few smaller and a few larger ones.

Box plots (pp. 63–70)

1

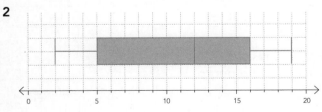

2

Understanding box plots (pp. 61–65)

1

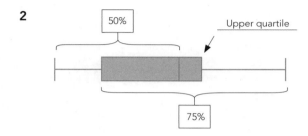

2

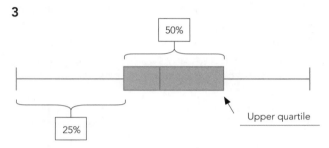

3

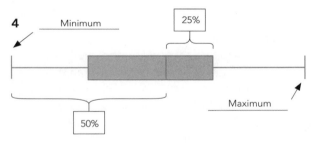

4

Comparing box plots (pp. 66–67)

1	a	✓		2	a	✗
	b	✗			b	✓
	c	✓			c	✗
	d	✓			d	✓
3	a	✓		4	a	✗
	b	✗			b	✓
	c	✗			c	✓
	d	✓			d	✗
5	a	✓			b	✓
	c	✗			d	✗

Describing meanings of features (pp. 68–70)

1 a **Centre**
     This means that they sell more chocolate ice creams on average than coconut ice creams.

   b **Spread**
     This means the number of chocolate ice creams sold is more variable than the number of coconut ice creams sold.

c **Shape**
Chocolate
I notice that the four sections of the box plot are reasonably similar in length.
This means that the distribution of chocolate ice cream sales is reasonably symmetrical and the sales are evenly spread.
Coconut
I notice that the two right sections of the box plot are smaller than the two on the left, so the data is skewed to the left.
This means that there is less variation in the numbers of ice creams sold when they sell a lot, but when there are fewer coconut ice creams sold, the numbers are more variable.

2 a **Centre**
I notice the median hourly pay rate for a student working in construction ($25) is smaller than the median pay rate for those working in IT ($28.50).
This means those working in IT on average have a higher hourly pay rate than those who work in construction.

b **Spread**
I notice the range for both the construction workers' and IT workers' hourly wage is $8.
This means there is a similar range of hourly wages paid in both industries.

c **Shape**
Construction
I notice the two sections on the left of the box plot for construction are smaller than those on the right, so the data is skewed to the right.
This means there is less variation in the lower hourly wage rates than in the higher hourly wage rates in construction.
IT
I notice that the four sections of the box plot for IT hourly wages are similar in length.
This means that the distribution in hourly wages for IT workers is reasonably symmetrical and evenly spread.

Making the call (pp. 71–75)

1 **I notice:** The medians of both samples lie outside the other's box.
This means:
☑ We can conclude that the individuals in population A tend to be bigger than those in population B.

2 **I notice:** The DBM is/is not bigger than ⅓ of the OVS.
This means: We can/cannot conclude that the individuals in population C tend to be bigger than those in population D.

3 **I notice:** The DBM is not bigger than ⅓ of the OVS.

This means: We cannot conclude that the individuals in population E tend to be bigger than those in population F.

4 **I notice:** The DBM is not bigger than ⅕ of the OVS.
This means: We cannot conclude that the individuals in population G tend to be bigger than those in population H.

5 **I notice:** The median of I lies below three quarters of J.
This means: We can conclude that the individuals in population J tend to be bigger than those in population I.

6 **I notice:** The DBM is bigger than ⅕ of the OVS.
This means: We can conclude that the individuals in population L tend to be bigger than those in population K.

7 **I notice:** The DBM is bigger than ⅕ of the OVS.
This means: We can conclude that Year 11s at Paradise High tend to take longer to get to school than Year 13s at Paradise High.

8 **I notice:** The DBM is not bigger than ⅓ of the OVS.
This means: We cannot conclude that the big dogs in Latisha's neighbourhood tend to go on longer walks than small dogs in her neighbourhood.

9 **I notice:** The DBM is not bigger than ⅕ of the OVS.
This means: We cannot conclude that EZ Radio tend to have longer ads than Pop n Roll.

Dot plots with box plots (p. 76)

		Dot plot	Box plot
		✓ or ✗	✓ or ✗
Centre	Mode	✓	✗
	Median	✓	✓
	Mean	✗	✗
Spread	Range	✓	✓
	Interquartile range	✗	✓
Shape		✓	✗
Symmetry		✓	✓
Unusual points and groupings		✓	✗
The value of every piece of data		✓	✗
Information for making the call		✗	✓

Practice task (pp. 80–81)
Question and purpose
Does the mass (g) of male whio tend to be greater than the mass of female whio from the Tasman area? Comparing the mass of male and female whio may help researchers to identify sex without having to disrupt and handle the birds more than necessary.

 ISBN: 9780170477437

Variation

Given that the data was collected for us, we must assume that variation was managed appropriately by the researchers. Here are some potential sources of variation:

- Measurement variation because the ducks were unlikely to have stayed still. A common technique to keep birds calm is to cover their heads with a breathable cloth. We assume the researchers were trained in appropriate bird-handling techniques to reduce stress and increase accuracy of measurement.
- Situation variation because we don't know how recently the birds have eaten. This could be managed by measuring birds at similar times of the day.

Centre

In the sample I notice that the median mass of male whio is 985 g and the median mass of female whio is 820 g. This is a difference of 165 g.

This means that in this sample the mass of the average male whio is larger than that of the average female whio.

Shape

In this sample the shape of the graph for female whio is bell shaped. The shape of that for male whio is also bell shaped apart from one mass measurement of 595 g. This male whio is 125 g less than the next male whio. This piece of data is unusual and may have come from a juvenile whio, or it may have been mistaken for a female whio, or it might be a measurement error.

Spread

The male whio masses (IQR = 150 g) are a little less spread out than those of the female whio masses (IQR = 156.5 g). This indicates that the female whio masses are slightly more varied than those of the male whio. While this difference is not large, it may be because some female whio are carrying eggs while others aren't.

Make the call

I notice that in this sample 50% of the male whio masses are larger than 50% of the female whio masses. This is only a small overlap of the boxes. This means that we can make the call that male whio from Tasman will tend to have larger masses on average than the female whio from the Tasman area. If I took another sample, it's likely that I would get different results. However, I am confident that I would come to the same conclusion.

Reflections

- The data would be more reliable if I had a bigger sample.
- It could be interesting to compare whio in Tasman to whio in other areas of New Zealand.

- Further investigation could be done on juvenile whio and compare them to adult whio.
- I have assumed that the sample is representative of the population of whio in the Tasman area, that the birds have been measured consistently, and the whio sex has been correctly identified and recorded.

Time series (pp. 82–111)
Features of time series (pp. 85–101)
1 The trend (pp. 85–92)

1 Stable   2 Inconsistent
3 Increasing   4 Decreasing
5 a Overall, the trend is ~~increasing/decreasing/~~ ~~stable/inconsistent~~ from the start of **2018** to the end of **2022**.
  b decreasing slightly
  c increasing slightly
6 a Overall, the trend has **decreased** from the start of **2010** to the end of **2019**.
  b No. Because it decreases quickly, is stable for a period of time and then increases most recently to just below the first value in 2010.
  c decreasing
  d mid-2013 until late 2015
  e increasing
7 a Overall, the trend has **decreased** from the start of **2010** to the end of **2019**.
  b mid-2010 and early 2011
  c The gradient is the steepest.
8 a **I notice** that overall, the trend has **increased** an average of **24 700 units** per quarter at the start of 2017 to **39 900 units** per quarter at the end of 2022.
  b **This means** that over this period, the **increase** is **39 900 – 24 700 = 15 200**.
  c **This represents** an average **increase** of **2530 (3 sf)** per year.
9 a **I notice** that overall, the trend has **decreased** from an average of **10 500 units** per quarter at the start of 2016 to **4630 units** per quarter at the end of 2022.
  b **This means** that over this period, the **decrease** is **10 500 – 4630 = 5870**.
  c **This represents** an average **decrease** of **838.57 (2 dp)** per year.
10 a **I notice** that overall, there has been a/an **decreasing** trend from an average of **28 600 units** per quarter at the start of 2018 to **26 400 units** per quarter at the end of 2022.
  b **This means** that over this period there has been a decrease of 28 600 – 26 400 = 2200.
  c **This represents** an average decrease of 440 per year.
11 a **I notice** that overall, the trend for the average number of earthquakes of strength 4 or more

per quarter in New Zealand each year has been **increasing**, from **224** in 2006 to **685** in 2023.

b **This means** that over this period, the number of earthquakes has **increased** by **461**.

c **This represents** an average **increase** of **27 (27.11)** per year.

12 a **I notice** that overall, there has been a decreasing trend in the average number of pairs of twins born per quarter in New Zealand from 864 at the start of 2009 to 753 at the end of 2022.

b **This means** that over this period there has been a decrease of 111 pairs per year.

c **This represents** an average decrease of 8 (8.54) pairs of twins per year.

13 a **I notice** that overall, there has been a decreasing trend in the average price per kg of avocados per month in New Zealand from $9.51 at the start of 2021 to $8.07 at the end of 2023.

b **This means** that over this period there has been a decrease in the price of avocados of $1.44 per year.

c **This represents** an average decrease in the price of avocados of 48 cents per year.

2 Repeating patterns (pp. 93–96)

1 a quarters   b yearly
  c 3   d 1

2 a months   b yearly
  c July/August d November/December

3 a **I notice** there is a recurring **yearly** pattern with a peak of about **800 hours** in **the first quarter** and a minimum of about **440 hours** in **the third quarter**.

b **This means** the highest number of sunshine hours in Tauranga occur in **January, February and March** and the lowest in **July, August and September**.

c Weather is generally better in summer, which includes January and February, and worst in winter, which includes July and August.

4 a **I notice** there is a recurring yearly pattern with a peak of about $18 to $21 in June or July and a minimum of about $5 to $6 in December or January.

b **This means** cucumbers are most expensive in the winter and least expensive in the summer.

c Cucumbers need warmth to grow, so they will be more abundant in summer than in winter.

5 **I notice** there is a recurring weekly pattern with a peaks of about 120 to 170 visitors in the weekends and a minimum of about 40 visitors on Mondays. **This means** most visitors come to the arcades in the weekends, and fewer on weekdays. The reason is that schools and many businesses are closed in the weekend, so this is when people have time to visit arcades.

3 Variation (pp. 97–99)

1 Decreasing   2 Constant
3 Increasing   4 Irregular

5 a **I notice** that the variation in the data is **increasing**.

b **This means** that the data is **becoming more variable**.

6 a **I notice** that the variation in the data between the start of 2015 until middle of 2021 is **consistent**, but after that it **increases**, and becomes consistent.

b **This means** that variation in the amount of wine exported from New Zealand was consistent between the start of 2015 until middle of 2021. After that the variation increased and almost doubled, but remained regular.

4 Unusual points (pp. 100–101)

1 **I notice** there is an unusually high point in the tenth month of 2023. It is a thousand larger than any other value.

2 **I notice** that the first quarter of 2022 is lower than we could have expected. All other quarter 1 values have been around 32, but this one is around 24.

3 **I notice** there is an unusually low value in the fourth month of 2022, approximately 40 000 lower than any other value. There is an unusually high value in the ninth quarter of 2022, around 50 000 higher than any other value.

Forecast or prediction (pp. 102–105)

1 a   2024Q1 **1775**   2024Q2 **1745**
     2024Q3 **1750**   2024Q4 **1700**

b
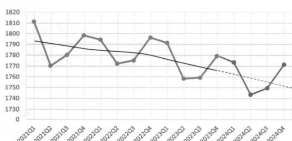

2 a   Week4D1 **78**   Week4D2 **62**
     Week4D3 **80**   Week4D4 **44**
     Week4D5 **97**   Week4D6 **180**
     Week4D7 **138**

b
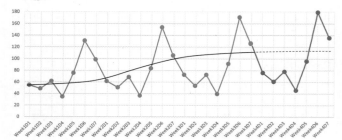

3 a 2024Q1 **15.8**   2024Q2 **13.7**
2024Q3 **8.1**   2024Q4 **12**

b

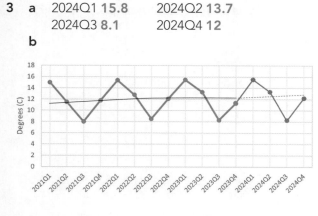

4 a 2024Q1 **610**   2024Q2 **490**
2024Q3 **300**   2024Q4 **560**

b

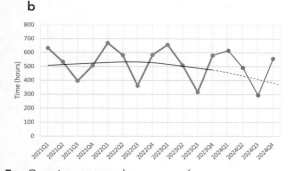

5	Consistent trend	✓
	Consistent pattern	✓
	Consistent variation	✓
	Enough data	✗
	Accurate predictions?	No
6	Consistent trend	✓
	Consistent pattern	✓
	Consistent variation	✓
	Enough data	✓
	Accurate predictions?	Yes
7	Consistent trend	✓
	Consistent pattern	✗
	Consistent variation	✓
	Enough data	✓
	Accurate predictions?	No
8	Consistent trend	✓
	Consistent pattern	✗
	Consistent variation	✓
	Enough data	✓
	Accurate predictions?	No

Practice task (pp. 109–111)

Question and purpose

Tracking cyclists through Hagley Park will help the planners at the council get a snapshot of the cycle use in the city. More specifically, how many cyclists are using this path in North Hagley Park and on which days of the week is it used the most?

Variation

Given that the data was collected for us, we must assume that variation was managed appropriately by the researchers. Here are some potential sources of variation:

- Measurement variation — we assume that the counter only registers cyclists and ignores skateboards, scooters and pedestrians. We also assume that if two cyclists are side by side, both are counted.
- Sampling variation — assume all cyclists on this path are counted and can't bypass or cycle around the counter. If so, then the sample might not be representative of the population.

Trend

I notice that there is an overall increase in the trend from 880 cyclists per day at the start of July to 959 cyclists in early August.

This means that over the 34 days there has been an increase of 79 cyclists. The average increase in the number of cyclists is 2 (2.32 to 2 dp) per day.

Pattern

I notice that there is a recurring weekly pattern in the data where there are more cyclists using this path on weekdays (at least 800) and far fewer (never more than 600) during the weekends.

This means significantly more cyclists are travelling through Hagley Park on weekdays than during the weekend. This suggests that this path is being used for the commute from suburbs into the town centre for work or school.

Variation in data

The variation is fairly stable, although reduces in the first week of August.

Unusual points

I notice that there is an unusually low number of cyclists passing the counter on Friday, 14 July. I have investigated this and discovered that it was Matariki — a public holiday. The number of cyclists on this day was more like the number expected on a day in the weekend.

There was higher numbers of cyclists than usual using the path between 17 and 19 July. There doesn't seem to be an obvious reason for this, but it is likely that the weather was warmer than usual during these days.

Predictions

In the second week of August I would predict that there will be around 600 cyclists during the weekend. On Monday, Tuesday and Wednesday, there will probably be just under 1600 cyclists, dropping to 1200 and 1300 on Thursday and Friday, respectively. I cannot be very confident about these predictions, because I think the number of cyclists, particularly during the weekends, will be partly dependent on the weather.

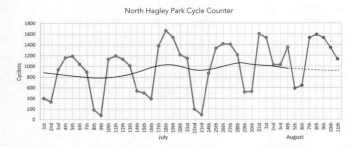

North Hagley Park Cycle Counter

Conclusions

Given there is a slight increase in trend and a consistent weekly pattern, I think the Christchurch City Council can be reasonably confident in these predictions for the days when the weather is reasonable. This information can help them plan for cycle path use through Hagley Park in the future.

Reflections

We are assuming that the cycle counters only count cyclists and not runners or scooter users. This may not be the case if the counters are based on mass. To be more confident in our conclusions, we could analyse the data over a longer period of time. This could include looking at patterns from year to year. You could also compare this data to weather and see if there is any relationship between poor weather and a reduced number of cyclists.

Experimental probability (pp. 112–159)
Probability (p. 112)

0	impossible, no chance, no way
0.1	very unlikely, slight chance, possible
0.2	improbable, possible, unlikely
0.3	improbable, maybe, possible, unlikely
0.4	maybe, possible, unlikely
0.5	maybe, possible, even chance, fifty-fifty
0.6	likely, probable
0.7	likely, probable
0.8	very likely, probable
0.9	very likely, extremely likely
1.0	definite, a sure thing, certain

Using numbers for writing probabilities (p. 113)

1 0.375, 0.4

More likely: $\frac{2}{5}$

2 $0.\dot{6}$, $0.\dot{6}\dot{3}$

More likely: $\frac{2}{3}$

3 0.85, 0.88

More likely: $\frac{22}{25}$

4 $0.2\dot{7}$, 0.25

More likely: $\frac{5}{18}$

5 0.125, $0.1\dot{3}$

Most likely: $\frac{2}{15}$

6 0.4375, $0.\dot{4}$

Most likely: $\frac{4}{9}$

7 Probabilities cannot be negative. ✗

8 $0.\dot{6}$ is between 0 and 1, so it's a valid probability ✓

9 0.70 is between 0 and 1, so it's a valid probability ✓

10 Probabilities cannot be greater than 1 or 100%. ✗

Ways of calculating probabilities (pp. 114–117)
1 Equally likely outcomes (pp. 114–115)

1 a $\frac{1}{6}$ or $0.1\dot{6}$ **b** $\frac{3}{6}$ or 0.5

  c $P(1, 3 \text{ or } 5) = \frac{3}{6}$ or 0.5

  d $P(1, 2, 4, 5 \text{ or } 6) = \frac{5}{6}$ or $0.8\dot{3}$

  e $P(1, 3, 5 \text{ or } 6) = \frac{4}{6}, \frac{2}{3}$ or $0.\dot{6}$

  f 0   **g** 1

2 a $\frac{1}{6}$ or $0.1\dot{6}$ **b** $\frac{2}{6}, \frac{1}{3}$ or $0.\dot{3}$

  c $P(2, 3 \text{ or } 5) = \frac{3}{6}$ or 0.5

  d $P(1, 4 \text{ or } 6) = \frac{3}{6}$ or 0.5

  e 0

  f $P(1, 2, 3 \text{ or } 5) = \frac{4}{6}, \frac{2}{3}$ or $0.\dot{6}$

  g $P(3 \text{ or } 5) = \frac{2}{6}, \frac{1}{3}$ or $0.\dot{3}$

3 a $\frac{5}{20}$ or $\frac{1}{4}$ or 0.25   **b** $\frac{15}{20}$ or $\frac{3}{4}$ or 0.75

  c $\frac{13}{20}$ or 0.65   **d** $\frac{7}{20}$ or 0.35

  e $\frac{14}{20}$ or 0.7   **f** $\frac{14}{20}$ or 0.7

  g 0   **h** 1

4 a $\frac{1}{28}$ **b** $\frac{7}{28}$

  c $\frac{2}{28}$ or $\frac{1}{14}$ **d** $\frac{3}{28}$

  e $\frac{4}{28}$ or $\frac{2}{14}$

2 Experimental probability (pp. 116–117)

1 $P(\text{no lunch}) = 0$

$P(\text{lunch from home}) = \frac{26}{48}$ or $\frac{13}{24}$ or 0.54

$P(\text{lunch from the canteen}) = \frac{13}{48}$ or 0.27

$P(\text{lunch from the dairy}) = \frac{9}{48} = \frac{3}{16}$ or 0.19

2 $P(\text{rugby}) = \frac{10}{28} = \frac{5}{14}$ or 0.36

$P(\text{basketball}) = \frac{7}{28}$ or 0.25

$P(\text{hockey}) = \frac{4}{28} = \frac{1}{7}$ or 0.14

$P(\text{netball}) = \frac{3}{28}$ or 0.11

$P(\text{badminton}) = \frac{3}{28}$ or 0.11

$P(\text{curling}) = \frac{1}{28}$ or 0.04

3 a $P(\text{none}) = \frac{32}{160}$ or 0.2

  b $P(0 < \text{time} \le 1) = \frac{83}{160}$ or 0.52

  c $P(\text{at least 2 hours}) = \frac{57}{160}$ or 0.36

4 a P(a pet) = $\frac{123}{160}$ or 0.77

b P(2 or 3 pets) = $\frac{55}{160}$ or 0.34

c P(1 pet) = $\frac{49}{160}$ × 100 = 31%

5 a 31

b P(walked) = $\frac{58}{160}$ or 0.36

c P(car) = $\frac{38}{160}$ or 0.24

Expected number of outcomes (p. 118)

1 25   **2** 26 or 27

3 65 or 66   **4** 27 or 28

5 1   **6** None

Probabilities from bar graphs (pp. 119–121)

1 a 20   **b** 0.4

c 70%   **d** 44

2 a 62   **b** 29.0%

c 0.919   **d** 108 lambs

3 a 80   **b** $\frac{16}{80}$ = 0.2

c $\frac{1}{6}$ or 0.$\dot{6}$

d She tossed the die only 80 times, so by chance she may have got more fours than she should have. If she performed the experiment a lot more times, then, provided the die was fair, she could expect the probability to be closer to 0.1$\dot{6}$.

e 20

f It would be similar with most bars between 11 and 15. Which die numbers have the lowest and highest bars cannot be predicted.

g The heights of the bars should be more uniform, with heights of about 167 units.

Probabilities with two events (pp. 122–131)

1 With equally likely outcomes (pp. 122–123)

1 a

	1	2	3	4	5	6
1	1 1	1 2	1 3	1 4	1 5	1 6
2	2 1	2 2	2 3	2 4	2 5	2 6
3	3 1	3 2	3 3	3 4	3 5	3 6
4	4 1	4 2	4 3	4 4	4 5	4 6
5	5 1	5 2	5 3	5 4	5 5	5 6
6	6 1	6 2	6 3	6 4	6 5	6 6

b 36   **c** (5 6) and (6 5)

d $\frac{2}{36}$ or $\frac{1}{18}$ or 0.0$\dot{5}$ **e** $\frac{6}{36}$ or $\frac{1}{6}$ or 0.1$\dot{6}$

f $\frac{8}{36}$ or $\frac{2}{9}$ or 0.$\dot{2}$ **g** $\frac{3}{36}$ or $\frac{1}{12}$ or 0.083$\dot{3}$

g $\frac{33}{36}$ or $\frac{11}{12}$ or 0.916$\dot{6}$ **i** $\frac{18}{36}$ or $\frac{1}{2}$ or 0.5

j $\frac{6}{36}$ or $\frac{1}{6}$ or 0.1$\dot{6}$ **k** $\frac{15}{36}$ or $\frac{5}{12}$ or 0.416$\dot{6}$

2 a

	1	2	3	4	5	6
1	1 1	1 2	1 3	1 4	1 5	1 6
2	2 1	2 2	2 3	2 4	2 5	2 6
4	4 1	4 2	4 3	4 4	4 5	4 6
8	8 1	8 2	8 3	8 4	8 5	8 6

b $\frac{1}{24}$ or 0.041$\dot{6}$ **c** $\frac{3}{24}$ or 0.125

d $\frac{12}{24}$ or $\frac{1}{2}$ or 0.5   **e** $\frac{5}{24}$ or 0.2083$\dot{3}$

3 a

	1	2	3	4	5	6	7	8
H	H 1	H 2	H 3	H 4	H 5	H 6	H 7	H 8
T	T 1	T 2	T 3	T 4	T 5	T 6	T 7	T 8

b $\frac{2}{16}$ or $\frac{1}{8}$ or 0.125   **c** $\frac{1}{16}$ or 0.0625

d $\frac{9}{16}$ or 0.5625   **e** $\frac{4}{16}$ or $\frac{1}{4}$ or 0.25

2 When probabilities of events are known (p. 124)

1 a $\frac{10}{20}$ or 0.5   **b** $\frac{11}{20}$ or 0.55

c $\frac{24}{400}$ or $\frac{3}{50}$ or 0.06   **d** $\frac{81}{400}$ or 0.2025

e $\frac{4}{380}$ or $\frac{1}{95}$ or 0.01   **f** $\frac{12}{380}$ or $\frac{3}{95}$ or 0.03

g $\frac{504}{6840}$ or 0.07   **h** 0

3 Using probability trees (pp. 125–128)

1 a

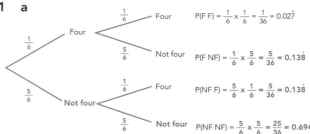

b $\frac{1}{6}$

c P(F NF) = $\frac{1}{6}$ × $\frac{5}{6}$

     = $\frac{5}{36}$ or 0.138$\dot{}$

d P(F F) = $\frac{1}{6}$ × $\frac{1}{6}$

     = $\frac{1}{36}$ or 0.027$\dot{}$

e P(F NF) **or** P(NF F) = $\frac{1}{6}$ × $\frac{5}{6}$ + $\frac{5}{6}$ × $\frac{1}{6}$

     = $\frac{5}{36}$ + $\frac{5}{36}$

     = $\frac{10}{36}$ or 0.27$\dot{}$

f $P(NF\ NF) = \frac{5}{6} \times \frac{5}{6}$

$= \frac{25}{36}$ or $0.69\dot{4}$

g $P(F\ NF)$ **or** $P(NF\ F)$ **or** $P(F\ F)$

$= \frac{1}{6} \times \frac{5}{6} + \frac{5}{6} \times \frac{1}{6} + \frac{1}{6} \times \frac{1}{6}$

$= \frac{5}{36} + \frac{5}{36} + \frac{1}{36}$

$= \frac{11}{36}$ or $0.30\dot{5}$

h They add to one because you must get either a four or not a four in the first two throws.

2 a

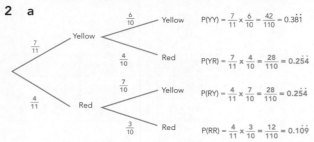

$P(YY) = \frac{7}{11} \times \frac{6}{10} = \frac{42}{110} = 0.38\dot{1}$

$P(YR) = \frac{7}{11} \times \frac{4}{10} = \frac{28}{110} = 0.25\dot{4}$

$P(RY) = \frac{4}{11} \times \frac{7}{10} = \frac{28}{110} = 0.25\dot{4}$

$P(RR) = \frac{4}{11} \times \frac{3}{10} = \frac{12}{110} = 0.10\dot{9}$

b $P(YY) = \frac{7}{11} \times \frac{6}{10}$

$= \frac{42}{110}$ or $0.38\dot{1}$

c $P(YR)$ **or** $P(RY) = \frac{7}{11} \times \frac{4}{10} + \frac{4}{11} \times \frac{7}{10}$

$= \frac{28}{110} + \frac{28}{110}$

$= \frac{56}{110}$ or $0.50\dot{9}$

d $P(YY)$ **or** $P(RR) = \frac{7}{11} \times \frac{6}{10} + \frac{4}{11} \times \frac{3}{10}$

$= \frac{42}{110} + \frac{12}{110}$

$= \frac{54}{110}$ or $0.49\dot{0}$

e $P(RR) = \frac{4}{11} \times \frac{3}{10}$

$= \frac{12}{110}$ or $0.10\dot{9}$

f $P(YR)$ **or** $P(RY)$ **or** $P(RR)$

$= \frac{7}{11} \times \frac{4}{10} + \frac{4}{11} \times \frac{7}{10} + \frac{4}{11} \times \frac{3}{10}$

$= \frac{28}{110} + \frac{28}{110} + \frac{12}{110}$

$= \frac{68}{110}$ or $0.61\dot{8}$

g $P(RR) = \frac{4}{11} \times \frac{3}{10}$

$= \frac{12}{110}$ or $0.10\dot{9}$

If she selects two red marbles, there will be two left in the bag.

h $P(RR) = \frac{12}{110}$

$\frac{12}{110} \times 200 = 21.8\dot{1}$

She would expect to get two red marbles 21 or 22 times.

4 Using two-way tables (pp. 129–131)

1

	Year 12	Year 13	Totals
Went to the Formal	84	96	180
Didn't go to the Formal	30	6	36
Totals	114	102	216

a $\frac{180}{216}$ or $0.8\dot{3}$ **b** $\frac{84}{216}$ or $0.3\dot{8}$

c $\frac{84}{114}$ or 0.74 **d** $\frac{180}{216} \times \frac{179}{215} = 0.69$

2

	Been snowboarding	Had not been snowboarding	Totals
Year 11	12	39	51
Year 9	9	36	45
Totals	21	75	96

a $\frac{75}{96}$ or 0.78 **b** $\frac{12}{96}$ or 0.125

c $\frac{9}{45}$ or 0.2 **d** $\frac{21}{96} \times \frac{20}{95} = 0.05$

e $\frac{39}{51} \times 443 = 338.76$

We would expect 338 or 339 Year 11 students from the school have not been snowboarding.

3

	First Class	Business Class	Economy	Totals
Noodles	7	8	186	201
Salad	11	12	59	82
Totals	18	20	245	283

a $\frac{82}{283} = 0.29$ **b** $\frac{7}{283} = 0.02$

c $\frac{12}{20} = 0.6$ **d** $\frac{59}{245} = 0.24$

e $\frac{38}{283} = 0.13$

$= 13\%$

4

	Fluoride	No fluoride	Totals
No tooth decay	358	196	554
Tooth decay	154	120	274
Totals	512	316	828

a $\frac{554}{828}$ or 0.67 **b** $\frac{196}{316}$ or 0.62

c $\frac{358}{512}$ or 0.70 **d** $\frac{274}{828} \times \frac{273}{827} = 0.11$

e $\frac{274}{828} \times 66\ 700 = 22\ 072.22$

We would expect 22 072 or 22 073 Year 8 students in New Zealand to have tooth decay.

Probability investigations (pp. 132–145)

2 What do we need to think about? Plan your investigation (pp. 133–136)

1

One trial	Outcomes	Event
Selecting two lollies.	• Selecting two lollies of the same colour. • Selecting two lollies of different colours.	Selecting two lollies of different colours.
Throwing the plane three times.	• Fewer than three planes travel at least 5 m. • All three paper aeroplanes travel at least 5 m.	All three paper planes travel at least 5 m.

2 Description of steps and recording of results

Liz will need to:

- put her hand in the bag, take a lolly out, note the colour and then give it to a classmate
- put her hand in the bag again and without looking take a second lolly out
- write down the colours of the first and second lolly selected (B for black or G for green).

Tane will have three throws of his paper plane. Tane needs to:

- mark a line that was exactly 5 m from the point at which he threw the plane
- throw the plane three times
- for each set of three throws, write down in a table the number that travelled at least 5 m before hitting the ground.

3 Sources of variation

Liz will need to:

- make sure the bag is shaken at the start, so that one colour isn't at the top
- make sure the bag is opaque so she cannot see what colour the lollies are
- make sure that all the lollies feel the same so she cannot feel for a particular colour.

Tane will need to:

- make sure the planes are all thrown from the same spot
- have somebody watching from the side to judge whether the plane has flown 5 m or not
- make sure the planes have not been damaged during a throw or landing
- make another plane that is as close as possible to the original if his plane does get damaged.

4 Number of trials

Liz will do 30 trials, one for every member of the class, including herself.

Tane will do 30 trials, each consisting of three attempts to throw a plane further than 5 m.

3 Record and display your results (pp. 137–140)

1 a Lollies

Frequencies

1st		B	G	Totals
	B	10	9	19
	G	6	5	11
	Totals	16	14	30

(2nd across the top)

Probabilities

	B	G	Totals
B	$\frac{10}{30}$ or 0.3	$\frac{9}{30}$ or 0.3	$\frac{19}{30}$ or 0.6\.3
G	$\frac{6}{30}$ or 0.2	$\frac{5}{30}$ or 0.1\.6	$\frac{11}{30}$ or 0.3\.6
Totals	$\frac{16}{30}$ or 0.5\.3	$\frac{14}{30}$ or 0.4\.6	1

b

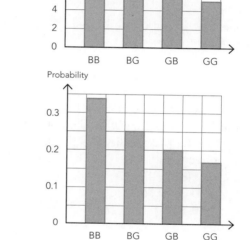

2 a Paper planes

Frequencies

	0	1	2	3	Total
Number that travelled 5 m	3	8	12	7	30

b Probabilities

	0	1	2	3	Total
Probability	$\frac{3}{30}$ or 0.1	$\frac{8}{30}$ or 0.2\.6	$\frac{12}{30}$ or 0.4	$\frac{7}{30}$ or 0.2\.3	1

b

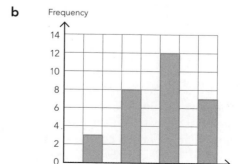

ISBN: 9780170477437

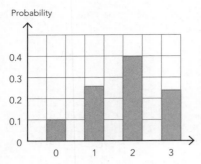

4 Describe what you see in your results and calculate the experimental probability (pp. 141–143)

1 Lollies

Frequencies

I notice the tallest bar is for selecting 2 black lollies (10) and the shortest is for selecting 2 green lollies (5).

This means that 15 out of the 30 pairs of lollies were the same colour.

Probabilities

I notice that the probability of selecting a black then a green was 0.3 and the probability of selecting a green then a black was 0.2.

This means half of the pairs were different colours.

The **experimental probability** that the lollies were the same colour

$$= P(BB) + P(GG)$$
$$= 0.\dot{3} + 0.1\dot{6}$$
$$= 0.5$$

2 Paper planes

Frequencies

I notice the tallest bar is for when his planes travelled 5 m or more in two out of three of his attempts.

This means that Tane was most likely to throw 5 m or more in two out of his three attempts.

Probabilities

I notice that the probability that his plane travelled 5 m or more in none of his three attempts is 0.1.

This means that his plane was unlikely to fail to reach 5 m in all three throws.

The **experimental probability** that Tane's plane flew further than 5 m in two or three of his attempts

$$= P(2) + P(3)$$
$$= 0.4 + 0.2\dot{3}$$
$$= 0.6\dot{3}$$

5 Writing a conclusion (pp. 144–145)

1 Lollies

The experimental probability that Liz selected 2 lollies of the same colour is 0.5.

My experimental probability seems reasonable because if she had 6 of each colour the probability of getting 2 the same would be 0.5, and having 5 of one colour and 7 of the other is not very different from 6 of each colour.

I could improve the reliability by doing a lot more trials.

I cannot think of any other factors apart from the variables mentioned earlier that might have affected my results.

Next steps: She could investigate what happens when there are, say, 4 of one colour and 8 of the other, or 3 and 9, etc.

2 Paper planes

The experimental probability that Tane's plane flew further than 5 m in two or three of his attempts is 0.6̇3̇.

He could improve the reliability by doing a lot more trials.

Another factor (other than the variables mentioned earlier) that might have affected his results is the location — was there any wind?

Next steps: He could investigate whether different designs of planes flew 5 m or more during three attempts.

Optional extras (pp. 146–152)

1 Plotting probabilites throughout an experiment (pp. 146–150)

1 Lollies

a

List of results	Total number of trials	Total number of trials where she selected two of the same colour	P(two the same)	
BB GB GB BG BB	5	2	$\frac{2}{5}$	0.4
GG BB GB BB GG	10	6	$\frac{6}{10}$	0.6
GG GB BG BG BG	15	7	$\frac{7}{15}$	0.4̇6̇
GB BG BG BB BB	20	9	$\frac{7}{20}$	0.45
GB BG BG BB BB	25	11	$\frac{11}{25}$	0.44
BG GG BB BB BB	30	15	$\frac{15}{30}$	0.5

b

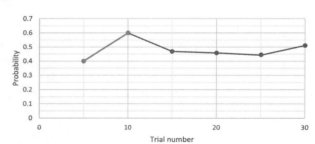

c The variation in the graph: the graph is unstable at the start but then stabilises between 15 and 20 trials, but goes up a bit at the end.

Towards the end, the graph stabilises at about between 0.45 and 0.5.

To be more certain of the probability of getting two lollies of the same colour, we would need to do more trials.

2 Paper planes

a

List of results	Total number of trials	Total number of times that Tane's plane flew further than 5 m in at least two attempts	P(at least two flew more than 5 m)	
3 3 1 2 2	5	4	$\frac{4}{5}$	0.8
1 0 1 2 2	10	6	$\frac{6}{10}$	0.6
0 2 2 2 3	15	10	$\frac{10}{15}$	$0.\dot{6}$
2 2 1 2 3	20	14	$\frac{14}{20}$	0.7
2 1 1 3 1	25	16	$\frac{16}{25}$	0.64
3 0 1 2 3	30	19	$\frac{19}{30}$	$0.6\dot{3}$

b

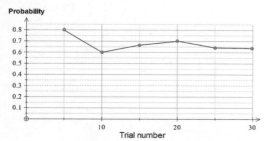

c The graph fluctuates at the start, but stabilises after 20 trials around a probability of about 0.64. In order to get a more accurate estimate of the probability, Tane would have to do more trials.

2 Calculating the theoretical probability (pp. 151–152)

1 Lollies

P(two lollies the same) = P(BB) + P(GG)

$$= \frac{7}{12} \times \frac{6}{11} \times \frac{5}{12} \times \frac{4}{11}$$

$$= \frac{42}{132} \times \frac{20}{132}$$

$$= \frac{62}{132}$$

$$= 0.41\dot{6}\dot{9}$$

My experimental probability was 0.5, which is higher than the theoretical probability. If I did more trials, I would expect to get an experimental probability that was closer to the theoretical probability.

2 Paper planes

We do not know the probability that a plane will travel 5 m or further, so we cannot calculate a theoretical probability.

Practice task (pp. 156–159)
Question and purpose

What is the probability that Huia will get a total of 7, 9 or a 12 when she throws two dice at the start of a Monopoly game? These are the numbers which would land her on a property that she could buy.

Plan your investigation

Outcomes and event:

Outcome 1: That she doesn't get a 6 or she gets a 6 and a total of 8, 10 or 11.

Outcome 2: She gets a 6 and a total of 7, 9 or 12.

Event of interest: She gets a 6 and a total of 7, 9 or 12.

List of the steps:

I will roll a pair of dice 50 times and record whether a 6 is rolled or not, and, if it is, the number on the other die. I will record my results in a table.

Sources of variation: I will assume that the dice are fair, and they are properly rolled so the numbers are random.

Variation

Measurement variation — incorrect recording of the results due to human error; all results should be checked by two people.

Induced variation — if the dice are not thoroughly shaken, you may not get random results.

Record, display and analysis of results

You may have done your investigation in a different way — if so, check with your teacher.

Table showing results, frequencies and probabilities:

- ✗ means no 6 at all.
- black numeral means 6 and a value that doesn't land me on a property.
- **green numeral** means a 6 and a value that does land me on a property.

List of results	Total number of trials	A 6 and a total of 7, 10 or 11	P(6 and a total of 7, 10 or 11)	
✗ ✗ ✗ ✗	5	0	$\frac{0}{5}$	0.8
✗ ✗ 6 2 4	10	1	$\frac{1}{10}$	0.6
1 2 ✗ ✗ 4	15	2	$\frac{2}{15}$	$0.1\dot{3}$
2 ✗ ✗ ✗ 2	20	2	$\frac{2}{20}$	0.2
✗ ✗ ✗ 3 ✗	25	3	$\frac{3}{25}$	0.12
✗ ✗ 2 1 ✗	30	4	$\frac{4}{30}$	$0.1\dot{3}$
✗ ✗ 5 ✗ ✗	35	4	$\frac{4}{35}$	0.114
3 2 ✗ ✗ 6	40	6	$\frac{6}{40}$	0.15
✗ ✗ ✗ ✗	45	6	$\frac{6}{45}$	$0.1\dot{3}$
✗ 2 ✗ ✗ 1	50	7	$\frac{7}{50}$	0.14

Frequency graph:

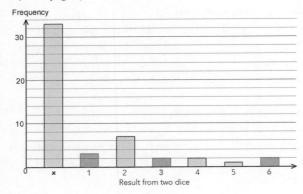

Probability graph:

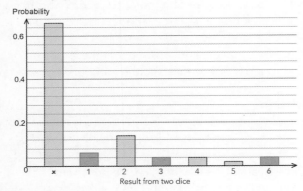

Optional extras

Probabilities throughout the experiment:

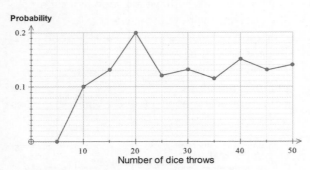

The graph shows that the probability of landing on a property varies between 0 and 0.2 during the first 20 throws, but it becomes more consistent as more trials are done, finishing up at 0.14.

Theoretical probability:
Die 1, die 2.

	1	2	3	4	5	6
1	1 1	1 2	1 3	1 4	1 5	(1 6)
2	2 1	2 2	2 3	2 4	2 5	2 6
3	3 1	3 2	3 3	3 4	3 5	(3 6)
4	4 1	4 2	4 3	4 4	4 5	4 6
5	5 1	5 2	5 3	5 4	5 5	5 6
6	(6 1)	6 2	(6 3)	6 4	6 5	(6 6)

Theoretical probability $= \dfrac{5}{36} = 0.13\dot{8}$

or

Theoretical probability $=$ P(**6**) x P(1 or 3) $+$
$\qquad\qquad\qquad$ P(6) x P(**1 or 3**) + P(6 and **6**)

$\qquad = \dfrac{1}{6} \times \dfrac{1}{3} + \dfrac{1}{6} \times \dfrac{1}{3} + \dfrac{1}{6} \times \dfrac{1}{6}$

$\qquad = \dfrac{1}{18} + \dfrac{1}{18} + \dfrac{1}{36}$

$\qquad = \dfrac{5}{36} = 0.13\dot{8}$

This probability is a little higher than my experimental probability.

Write a conclusion

My experimental probability of getting a number that allows Aroha to buy a property was 0.14, very close to the theoretical probability of 0.13$\dot{8}$. In general the more trials that are completed, the closer the experimental probability and theoretical probability will become. A probability of 0.13$\dot{8}$ seems reasonable because throwing a number that allows Aroha to buy a property is pretty unlikely. I could extend this investigation by trying to find out the probabilities of buying 0, 1, 2 or 3 properties in the 2 throws of the dice.